2B

英语语境语法

(第四版)

Grammar in Context

4TH EDITION

SANDRA N. ELBAUM

北京大学出版社

PEKING UNIVERSITY PRESS

著作权合同登记　图字：01-2006-0573

图书在版编目(CIP)数据

英语语境语法. 2B / 桑德拉编著. —北京：北京大学出版社，2006.1
(英语语境语法系列丛书)
ISBN 7-301-10311-5

Ⅰ. 英…　Ⅱ. 桑…　Ⅲ. 英语-语法　Ⅳ. H314

中国版本图书馆 CIP 数据核字(2005)第 145677 号

SANDRA N. ELBAUM
Grammar in Context 2B, Fourth Edition
EISBN: 1-4130-0744-9

书　　　名：英语语境语法　2B　(第四版)
著作责任者：SANDRA N. ELBAUM　编著
责 任 编 辑：胡　娜　张　冰
标 准 书 号：ISBN 7-301-10311-5/H·1604
出 版 发 行：北京大学出版社
地　　　址：北京市海淀区成府路 205 号　100871
网　　　址：http://cbs.pku.edu.cn
电　　　话：邮购部 62752015　发行部 62750672　编辑部 62767347
电 子 邮 箱：zbing@pup.pku.edu.cn
排 版 者：华伦图文制作中心
印 刷 者：北京大学印刷厂
经 销 者：新华书店
　　　　　889毫米×1194毫米　16 开本　18.5 印张　510 千字
　　　　　2006 年 1 月第 1 版　2006 年 1 月第 1 次印刷
定　　　价：32.00 元 (配有光盘)

导　言

北京大学英语系教授　王逢鑫

一

　　语言由语音、词汇和语法三个要素组成。学习一门外语,必须掌握这三个要素,缺一不可。有人认为只要记住单词,能读出音来,就行了,而语法可有可无。其实这是一种误解。语法是组词造句的法则,十分重要。传统英语语法细分为词法(morphology)和句法(syntax)。词法解释词分为哪些种类,即词类;告诉人们每个词类有什么特点,即词性;说明一个词与别的什么词可以联系在一起使用,即在句子里起什么作用。英语词汇形态与汉语有很大的区别。例如,名词有单、复数之分,还有可数与不可数之分。人称代词有主格、宾格和所有格之分。动词有现在式、过去式和过去分词三种不同形式;还有不定式、现在分词、过去分词和动名词等非谓语动词形式。形容词有原级、比较级和最高级三种形式。数词有基数词和序数词之分。以上词类大都是规则变化,但是也有很多不规则变化的例外情况。例如英语有一百来个不规则动词,其中多数是常用动词。介词后面跟人称代词要用宾格,跟动词要用动名词形式。英语的冠词更是难学。有人学了多年英语,还是弄不清楚什么时候用定冠词,什么时候用不定冠词,什么时候不用任何冠词。虽然不定冠词仅有 a 和 an 两种形式, 但是有人把 an hour 写成 a hour, 把 a university 写成 an university。这些繁杂的内容都是初学者必须掌握的,使用不当就要犯错误。

　　英语句法分析句子的种类、结构和功能。英语句法比汉语复杂。英语有各种各样的时态,每种时态有自己固定的形式,不能用错。句法规则繁多,几乎没有什么道理可讲。例如,在一般现在时里,单数第三人称的动词要加-s。情态动词和助动词后面要跟原形动词。英语句子讲究人称和时态前后呼应,左右照顾。诸如此类的条条框框都是初学者的"拦路虎"。

　　国内外的语言学家和英语教师,曾经尝试使用各种各样的方法来教英语语法。时代不同,学习目的不同,教学对象不同,教材不同,学习方法不同,使得人们很难找到学习英语语法的一个最佳方案。但是,我们了解一下国内外英语语法教学的来龙去脉,或许有助于我们吸取教训,总结经验,寻找有效的学习英语语法的途径。

　　传统法(traditional method)强调以语法为纲,以语法为教学中心。学生按部就班学习语法规则,先是死记硬背条条框框,然后做大量的机械性(mechanical)练习,基本上是没有上下文的单句翻译、语法填空和造句练习。追求的是语法形式正确无误,而不管在什么情况下使用语言。学习语法,不是为了交际,而是为了阅读内容艰深的文章,分析复杂的句法结构。我国解放前和解放初期的英语专业大学生,是通过传统法学习英语语法的,虽然有一些弊端,例如引导学生重视阅读和笔译,而忽视口头表达能力的培养。但是这种方法也并非一无是处。学生中不乏精通英语的成功者。传统法强调阅读小说、诗歌、戏剧和散文等文学作品,认为文学语言是最好的语言。通过对语句和篇章细致的句法分析,学生获得对语句和篇章结构,尤其是繁杂结构的精确理解。今天我们强调学习语言是为了交流思想,重视口头表达能力,传统法是难当此任的。

听说法(audio-visual method)将英语分成许多基本句型(sentence pattern),将语法教学与句型教学结合在一起。要求学生熟练掌握句型,反复口头练习,达到不假思索、脱口而出的程度。掌握了句型,就等于掌握了语法。20世纪60年代初,听说法引入我国,在当时的英语专业大学生中间曾经奏效。学生反复练习没有上下文的基本句型,虽然枯燥无味,但是在当时的历史环境下,多数学生能够不厌其烦地做大量的机械性口头练习,而取得较好的学习效果。现在的学生要求在学习过程中有更多的独立自主,对死记硬背基本句型不太感兴趣。利用听说法学习英语语法似乎不太合乎时宜了。

语言学家和英语教师总是想方设法改进语法教学。他们先是将以单句练习为主的机械性句型练习,扩充为共有两句话的二人对话,构成一个简单的情景,使所练习的句型变得有意义。再往后,进一步将二人对话扩大为围绕一个主题的、有上下文的情景会话(situational conversation)。这样,学生可以在一定的语境(context)之中通过句型学习英语语法。最初的语境是为了练习某个语言点,或为了掌握某种意念功能而编造的,具有人为的成分。20世纪80年代初,国内外兴起交际法(communicative method)。这种教学法的目标是让学生不仅学会听、说、读、写的语言能力 (linguistic competence),还要掌握交际能力(communicative competence)。交际法从交流的目的出发,既要求语法正确(correct in grammar),更要求语用得体(appropriate in use)。因而在教学中引进了社会与文化因素。学习内容不再是干巴巴的基本句型,而是人们关注的社会问题和文化现象。学生不再为学习语法而学习语法,而是为了交际来学习语法。他们希望能够使用语法正确、语用得体的语言,就人们关注的社会问题和文化现象进行交流。这样就需要在一定的社会环境和文化语境里学习语言,包括学习语法。在学习语言的同时,必须了解英语国家的文化背景, 以及中外文化差异。只有学习了相关的文化背景知识,才能更好地掌握语言。这套名为Grammar in Context (《英语语境语法》)的教材,在上述背景下应运而生。

二

这套《英语语境语法》的编者 Sandra N. Elbaum 女士,是美国的一位英语教师,专门教授从世界各地到美国的移民,他们是以英语为第二语言的学生。Elbaum 女士幼年随父母由波兰移民到美国,语言差异和文化差异经常使她的父母感到困惑。Elbaum 女士在移民聚居的社区中成长,深知一个外国移民在美国生存,不仅要逾越语言障碍,更要克服文化差异。她有一个信念,就是通过语境学习语法。她不但在教学中身体力行,通过语境教英语语法,而且亲自编写教材,体现这一理念。

这套英语教材名曰《英语语境语法》,实际上是教给学生通过语境学习英语语言。这套教材的宗旨是:让学习者在语境中学习语法,以便学到更多东西,记住更多东西,更加有效地运用语言。

这套教材有如下突出特点:

1. 教给学生进行口头交流和书面交流所必需的语法知识。按照循序渐进原则安排语法点,讲解后面的语法内容都联系和复习前面的语法内容,使整个语法系统构成一个有机的整体。解释每个语法点,都使用形象的语法图表(grammar chart),一目了然。每个语法图表提供有语境的精选例句,并给出清晰的解释,还配以语言提示(language note),增强学习者对所学语法结构的理解。每个语法点还以图表方式解释其形式、用途、语序、主语、相关结构、描述与定义、所需介词搭配、肯定句、否定句和疑问句及回答等项目。每个项目都配有大量的口头和笔头练习。

2. 不是为教语法而教语法，而是通过语法教学，给学习者提供有用的(useful)、有意义的(meaningful)技能和基本文化知识。在课堂上，教师不是局限于让学生做机械性练习，而是让他们通过二人对话、小组活动、游戏、讨论等多种形式的扩展活动(expansion activity)，互相启发，互相帮助，学以致用。通过阅读、作文、独立思考的练习等方式，学习者拓展自己的语言知识和交际能力，最终达到既能有效使用语言，又有信心正确使用语言进行交流的双重目的。

3. 教材将英语语法学习和美国文化语境结合起来。全套教材分为1、2、3三级，每级又分为两个分册，共有1A、1B、2A、2B、3A、3B等6个分册。1级和2级各有14课；3级有10课。每课内容，包括语法讲解和练习、阅读课文和扩展活动，都围绕美国社会的一个热门话题，构成一个语境。从语言学习角度，涉及一个语法点；从文化学习角度，涉及一个话题。二者巧妙结合。1级有学校生活、美国政府、美国节日、美国人及其住宅、家庭与姓名、美国人生活方式、婚礼、飞行、购物、营养与健康、伟大女性、美国地理、约会与婚姻、实习等14个话题。2级有宠物、老年生活、改善生活、婚礼、感恩节与印第安人、健康、移民、租房、上网搜索、找工作、交友、体育、法律、货币等14个话题。3级有工作、好莱坞、灾难与悲剧、消费者警告、肯尼迪家族、计算机与互联网、帮助他人、来到美国、关爱儿童、科学与科幻小说等10个话题。这些语境概括了美国社会的方方面面，是了解美国文化和在美国生存所必需的基本知识。这些语境有助于学习者掌握必需的文化背景知识，使他们懂得美国文化在语言、信仰和日常生活情景等方面的重要作用。

这是一套通过语境学习英语的好教材。使用这套教材，学习者不仅可以熟练掌握英语语法，运用英语语言；而且可以学习美国文化背景知识，在语境中学习英语，在语境中使用英语。希望学习者喜欢这套教材，并通过学习这套教材学好英语。

In loving memory of
Roberto Garrido Alfaro

Acknowledgements

Many thanks to Dennis Hogan, Jim Brown, Sherrise Roehr, Yeny Kim, and Sally Giangrande from Thomson Heinle for their ongoing support of the *Grammar in Context* series. I would especially like to thank my editor, Charlotte Sturdy, for her keen eye to detail and invaluable suggestions.

And many thanks to my students at Truman College, who have increased my understanding of my own language and taught me to see life from another point of view. By sharing their observations, questions, and life stories, they have enriched my life enormously—*Sandra N. Elbaum*

Thomson Heinle would like to thank the following people for their contributions:

Marki Alexander
Oklahoma State
 University
Stillwater, OK

Joan M. Amore
Triton College
River Grove, IL

**Edina Pingleton
Bagley**
Nassau Community
 College
Garden City, NY

Judith A. G. Benka
Normandale Community
 College
Bloomington, MN

**Judith Book-
Ehrlichman**
Bergen Community
 College
Paramus, NJ

Lyn Buchheit
Community College of
 Philadelphia
Philadelphia, PA

Charlotte M. Calobrisi
Northern Virginia
 Community College
Annandale, VA

Sarah A. Carpenter
Normandale Community
 College
Bloomington, MN

Jeanette Clement
Duquesne University
Pittsburgh, PA

Allis Cole
Shoreline Community
 College
Shoreline, WA

**Jacqueline M.
Cunningham**
Triton College
River Grove, IL

Lisa DePaoli
Sierra College
Rocklin, CA

Maha Edlbi
Sierra College
Rocklin, CA

Rhonda J. Farley
Cosumnes River College
Sacramento, CA

Jennifer Farnell
University of
 Connecticut
American Language
 Program
Stamford, CT

**Abigail-Marie
Fiattarone**
Mesa Community
 College
Mesa, AZ

Marcia Gethin-Jones
University of
 Connecticut
American Language
 Program
Storrs, CT

Linda Harlow
Santa Rosa Junior
 College
Santa Rosa, CA

Suha R. Hattab
Triton College
River Grove, IL

Bill Keniston
Normandale Community
 College

Bloomington, MN

Walton King
Arkansas State
 University
Jonesboro, AR

Kathleen Krokar
Truman College
Chicago, IL

John Larkin
NVCC-Community and
 Workforce
 Development
Annandale, VA

Michael Larsen
American River College
Sacramento, CA

Bea C. Lawn
Gavilan College
Gilroy, CA

Rob Lee
Pasadena City College
Pasadena, CA

**Oranit
Limmaneeprasert**
American River College
Sacramento, CA

Gennell Lockwood
Shoreline Community
 College
Shoreline, WA

Linda Louie
Highline Community
 College
Des Moines, WA

Melanie A. Majeski
Naugatuck Valley
 Community College
Waterbury, CT

Maria Marin
De Anza College
Cupertino, CA

Karen Miceli
Cosumnes River College
Sacramento, CA

Jeanie Pavichevich
Triton College
River Grove, IL

Herbert Pierson
St. John's University
New York City, NY

Dina Poggi
De Anza College
Cupertino, CA

Mark Rau
American River College
Sacramento, CA

John W. Roberts
Shoreline Community
 College
Shoreline, WA

Azize R. Ruttler
Bergen Community
 College
Paramus, NJ

Ann Salzmann
University of Illinois,
Urbana, IL

Eva Teagarden
Yuba College
Marysville, CA

Susan Wilson
San Jose City College
San Jose, CA

Martha Yeager-Tobar
Cerritos College
Norwalk, CA

Contents

Lesson 2   43

Lesson 3   87

Lesson 6   187

Lesson 7   211

Lesson 8   237

Lesson 9 **277**

Lesson 10 **315**

Lesson 11 353

Lesson 12 375

Lesson 13   413

Lesson 14   437

Appendices

A word from the author

It seems that I was born to be an ESL teacher. My parents immigrated to the U.S. from Poland as adults and were confused not only by the English language but by American culture as well. Born in the U.S., I often had the task as a child to explain the intricacies of the language and allay my parents' fears about the culture. It is no wonder to me that I became an ESL teacher, and later, an ESL writer who focuses on explanations of American culture in order to illustrate grammar. My life growing up in an immigrant neighborhood was very similar to the lives of my students, so I have a feel for what confuses them and what they need to know about American life.

ESL teachers often find themselves explaining confusing customs and providing practical information about life in the U.S. Often, teachers are a student's only source of information about American life. With **Grammar in Context, Fourth Edition**, I enjoy sharing my experiences with you.

Grammar in Context, Fourth Edition connects grammar with American cultural context, providing learners of English with a useful and meaningful skill and knowledge base. Students learn the grammar necessary to communicate verbally and in writing, and learn how American culture plays a role in language, beliefs, and everyday situations.

Enjoy the new edition of **Grammar in Context!**

Sandra N. Elbaum

Grammar in Context

Students learn more, remember more, and use language more effectively when they learn grammar in context.

Learning a language through meaningful themes and practicing it in a contextualized setting promote both linguistic and cognitive development. In **Grammar in Context,** grammar is presented in interesting and culturally informative readings, and the language and context are subsequently practiced throughout the chapter.

New to this edition:

- New and updated readings on current American topics such as Instant Messaging and eBay.
- Updated grammar charts that now include essential language notes.
- Updated exercises and activities that provide contextualized practice using a variety of exercise types, as well as additional practice for more difficult structures.
- New lower-level *Grammar in Context Basic* for beginning level students.
- New wrap-around Teacher's Annotated Edition with page-by-page, point-of-use teaching suggestions.
- Expanded Assessment CD-ROM with ExamView® Pro Test Generator now contains more questions types and assessment options to easily allow teachers to create tests and quizzes.

Distinctive Features of *Grammar in Context:*

Students prepare for academic assignments and everyday language tasks.

Discussions, readings, compositions, and exercises involving higherlevel critical thinking skills develop overall language and communication skills.

Students expand their knowledge of American topics and culture.

The readings in **Grammar in Context** help students gain insight into and enrich their knowledge of American culture and history. Students gain ample exposure to the practicalities of American life, such as writing a résumé, dealing with telemarketers and junk mail, and getting student internships. Their new knowledge helps them adapt to everyday life in the U.S.

Students learn to use their new skills to communicate.

The exercises and Expansion Activities in **Grammar in Context** help students learn English while practicing their writing and speaking skills. Students work together in pairs and groups to find more information about topics, to make presentations, to play games, and to role-play. Their confidence in using English increases, as does their ability to communicate effectively.

Welcome to **Grammar in Context, Fourth Edition**

Students learn more, remember more, and use language more effectively when they learn grammar in context.

Grammar in Context, Fourth Edition connects grammar with rich, American cultural context, providing learners of English with a useful and meaningful skill and knowledge base.

An Audio Program allows students to hear the readings and dialogs, and provides an opportunity to practice their listening skills.

Readings on American topics such as Google, Internet Matchmaking, and Jury Duty present and illustrate the grammatical structure in an informative and meaningful context.

Grammar charts offer clear explanations and provide contextualized examples of the structure.

Language Notes refine students' understanding of the target structure.

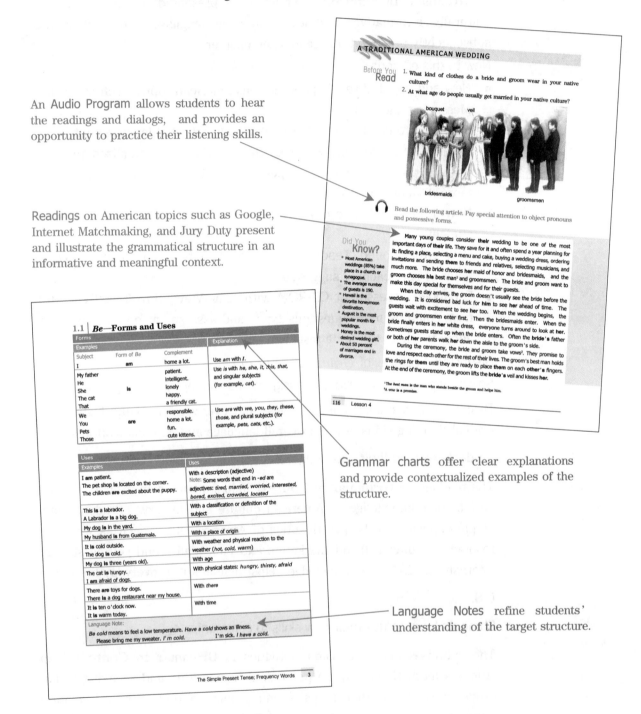

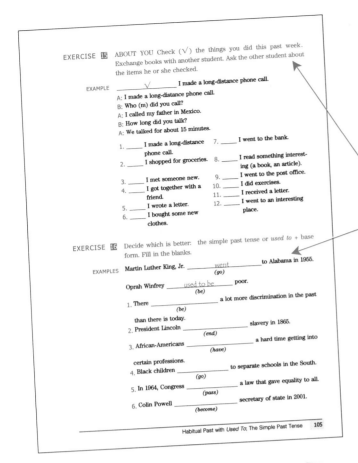

A variety of contextualized activities keeps the classroom lively and targets different learning styles.

A Summary provides the lesson's essential grammar in an easy-to-reference format.

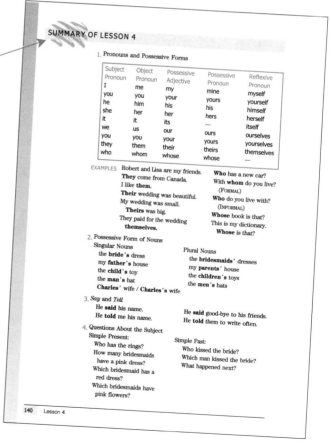

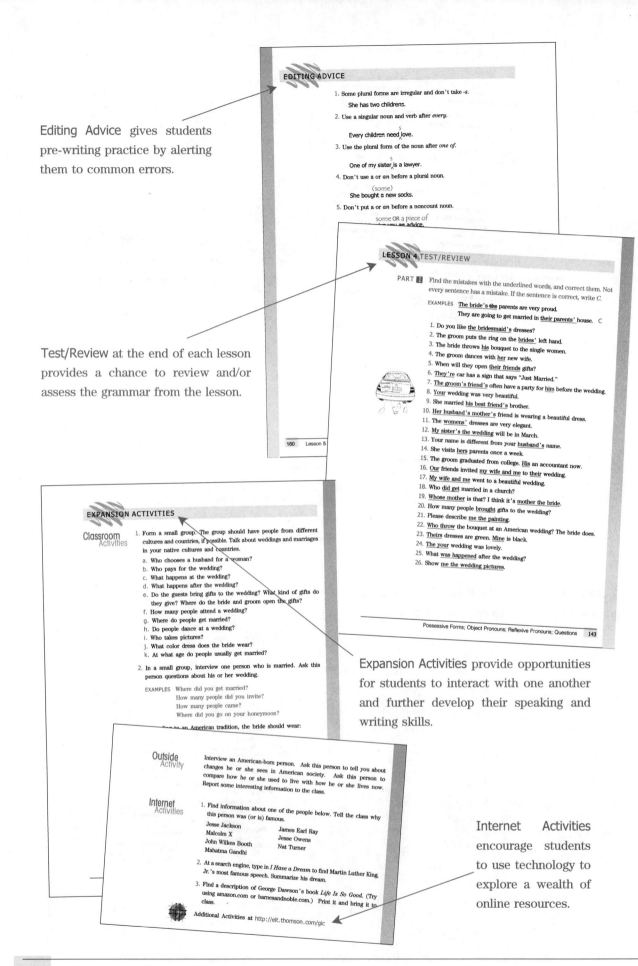

Editing Advice gives students pre-writing practice by alerting them to common errors.

EDITING ADVICE

1. Some plural forms are irregular and don't take -s.
   She has two childrens.

2. Use a singular noun and verb after *every*.
   Every children need love.

3. Use the plural form of the noun after *one of*.
   One of my sister is a lawyer.

4. Don't use a or *an* before a plural noun.
   (some)
   She bought a new socks.

5. Don't put a or *an* before a noncount noun.
   some OR a piece of

Test/Review at the end of each lesson provides a chance to review and/or assess the grammar from the lesson.

LESSON 4 TEST/REVIEW

PART **1**   Find the mistakes with the underlined words, and correct them. Not every sentence has a mistake. If the sentence is correct, write *C*.

EXAMPLES   The bride's ~~she~~ parents are very proud.
They are going to get married in their parents' house.   C

1. Do you like the bridesmaid's dresses?
2. The groom puts the ring on the brides' left hand.
3. The bride throws his bouquet to the single women.
4. The groom dances with her new wife.
5. When will they open their friends gifts?
6. They're car has a sign that says "Just Married."
7. The groom's friend's often have a party for him before the wedding.
8. Your wedding was very beautiful.
9. She married his best friend's brother.
10. Her husband's mother's friend is wearing a beautiful dress.
11. The womens' dresses are very elegant.
12. My sister's the wedding will be in March.
13. Your name is different from your husband's name.
14. She visits hers parents once a week.
15. The groom graduated from college. His an accountant now.
16. Our friends invited my wife and me to their wedding.
17. My wife and me went to a beautiful wedding.
18. Who did get married in a church?
19. Whose mother is that? I think it's mother the bride.
20. How many people brought gifts to the wedding?
21. Please describe me the painting.
22. Who throw the bouquet at an American wedding? The bride does.
23. Theirs dresses are green. Mine is black.
24. The your wedding was lovely.
25. What was happened after the wedding?
26. Show me the wedding pictures.

Possessive Forms; Object Pronouns; Reflexive Pronouns; Questions   143

EXPANSION ACTIVITIES

Classroom Activities

1. Form a small group. The group should have people from different cultures and countries, if possible. Talk about weddings and marriages in your native cultures and countries.
   a. Who chooses a husband for a woman?
   b. Who pays for the wedding?
   c. What happens at the wedding?
   d. What happens after the wedding?
   e. Do the guests bring gifts to the wedding? What kind of gifts do they give? Where do the bride and groom open the gifts?
   f. How many people attend a wedding?
   g. Where do people get married?
   h. Do people dance at a wedding?
   i. Who takes pictures?
   j. What color dress does the bride wear?
   k. At what age do people usually get married?

2. In a small group, interview one person who is married. Ask this person questions about his or her wedding.

   EXAMPLES   Where did you get married?
   How many people did you invite?
   How many people came?
   Where did you go on your honeymoon?

According to an American tradition, the bride should wear:

Expansion Activities provide opportunities for students to interact with one another and further develop their speaking and writing skills.

Outside Activity

Interview an American-born person. Ask this person to tell you about changes he or she sees in American society. Ask this person to compare how he or she used to live with how he or she lives now. Report some interesting information to the class.

Internet Activities

1. Find information about one of the people below. Tell the class why this person was (or is) famous.
   Jesse Jackson
   Malcolm X
   John Wilkes Booth
   Mahatma Gandhi
   James Earl Ray
   Jesse Owens
   Nat Turner

2. At a search engine, type in *I Have a Dream* to find Martin Luther King, Jr.'s most famous speech. Summarize his dream.

3. Find a description of George Dawson's book *Life Is So Good*. (Try using amazon.com or barnesandnoble.com.) Print it and bring it to class.

Additional Activities at http://elt.thomson.com/gic

Internet Activities encourage students to use technology to explore a wealth of online resources.

Grammar in Context Student Book Supplements

Audio Program
- Audio CDs and Audio Tapes allow students to listen to every reading in the book as well as selected dialogs.

More Grammar Practice Workbooks
- Can be used with *Grammar in Context* or any skills text to learn and review the essential grammar
- Great for in-class practice or homework
- Includes practice on all grammar points in *Grammar in Context*

Teacher's Annotated Edition
- New component offers page-by-page answers and teaching suggestions.

Assessment CD-ROM with ExamView® Pro Test Generator
- Test Generator allows teachers to create tests and quizzes quickly and easily.

Interactive CD-ROM
- CD-ROM allows for supplemental interactive practice on grammar points from *Grammar in Context*.

Split Editions
- Split editions provide options for short courses.

Instructional Video
- Video offers teaching suggestions and advice on how to use *Grammar in Context*.

Web Site
- Web site gives access to additional activities and promotes the use of the Internet.

Toolbox
- A WebTutor™ Toolbox available on WebCT™ or Blackboard® provides chapter-by-chapter quizzes and support.

LESSON

8

GRAMMAR

Modals
Related Expressions

CONTEXT: Renting an Apartment

An Apartment Lease
Tenants' Rights
The New Neighbors
At a Garage Sale

APARTMENT
FOR RENT

8.1 | Modals and Related Expressions—An Overview

A modal adds meaning to the verb that follows it.

List of Modals	Modals are different from other verbs in several ways.

can
could
should
will
would
may
might
must
	1. The base form of a verb follows a modal.[1]
	    You **must pay** your rent. (*Not:* You must <u>to</u> pay your rent.)
	    He **should clean** his apartment now. (*Not:* He should clean<u>ing</u>
	      His apartment now.)
	2. Modals never have an *-s, -ed,* or *-ing* ending.
	    He **can** rent an apartment. (*Not:* He can<u>s</u> rent an apartment.)

Related Expressions	Some verbs are like modals in meaning.
have to be able to be supposed to be permitted to be allowed to	He **must** sign the lease. = He **has to** sign the lease. He **can** pay the rent. = He **is able to** pay the rent. I **must** pay my rent by the first of the month. = I'm **supposed to** pay my rent by the first of the month. You **can't** change the locks in your apartment. = You **are not permitted** **to** change the locks in your apartment. = You **are not allowed to** change the locks in your apartment.

AN APARTMENT LEASE

Before You Read

1. Do you live in an apartment? Do you have a lease? Did you understand the lease when you signed it?

2. What kinds of things are not allowed in your apartment?

> # LEASE
>
> The following is a lease agreement (hereinafter referred as the "Agreement") legal and a r___ on is day: ____
> of this month: ____ of this year: ____ Between _____(hereinafter referred to as
> Lessor) and _____(hereinafter referred to as Lessee)
>
> This agreement is for the property of: _____ for the bimonthly amount of: _____ for the term
> of: _____ with this number of occupants: _____
>
> The lessor is the owner of the property and the Lessee is given permission to use the property as specified in
> the agreement and must comply with this agreement or will cause a breach of contract at that time releasing
> rights of this agreement and allowing the lessor to use any legal means to repair any property damage or loss.
>
> This agreement is also bound by the Lessor and any right of the Lessee that the Lessor doesn't comply with by
> this agreement at that time releases the tenant from the agreement and allows the Lessee to obtain numerous
> options to help the Lessee's rights be enforced and/or compensated.

[1] Do not follow a modal with an infinitive. There is one exception: *ought to. Ought to* means *should.*

Read the following article. Pay special attention to modals and related expressions.

When people rent an apartment, they often **have to** sign a lease. A lease is an agreement between the owner (landlord[2]) and the renter (tenant). A lease states the period of time for the rental, the amount of the rent, and the rules the renter **must** follow. Some leases contain the following rules:

- Renters **must not** have a waterbed.
- Renters **must not** have a pet.
- Renters **must not** change the locks without the owner's permission.
- Renters **must** pay a security deposit.

Many owners ask the renters to pay a security deposit, in case there are damages. When the renters move out, the owners **are supposed to** return the deposit plus interest if the apartment is in good condition. If there is damage, the owners **can** use part or all of the money to repair the damage. However, they **may not** keep the renters' money for normal wear and tear (the normal use of the apartment).

Renters **do not have to** agree to all the terms of the lease. They can ask for changes before they sign. A pet owner, for example, can ask for permission to have a pet by offering to pay a higher security deposit.

There are laws that protect renters. For example, owners **must** provide heat during the winter months. In most cities, they **must** put a smoke detector in each apartment and in the halls. In addition, owners **can't** refuse to rent to a person because of sex, race, religion, nationality, or disability.

When the lease is up for renewal, owners can offer the renters a new lease or they can ask the renters to leave. The owners **are supposed to** notify the renters (usually at least 30 days in advance) if they want the renters to leave.

smoke detector

[2] A *landlord* is a man. A *landlady* is a woman.

8.2 | Negatives with Modals

Negatives and negative contractions	You form the negative of a modal by putting *not* after the modal. You can make a negative contraction with some, but not all, modals.

cannot → can't
could not → couldn't
should not → shouldn't
will not → won't
would not → wouldn't
may not → (no contraction)
might not → (no contraction)
must not → mustn't
	• The negative of *can* is *cannot* (one word) or *can't*.
	    We **cannot** pay the rent.
	    You **can't** have a dog in your apartment.
	• The negative contraction of *will not* is *won't*.
	    We **will not** renew our lease.
	    We **won't** stay here.
	• Don't make a contraction for *may not* or *might not*.
	    You **may not** know legal terms.
	    You **might not** understand the lease.

EXERCISE **1**   Write the negative form of the underlined words. Use a contraction whenever possible.

EXAMPLE   You <u>must</u> pay a security deposit. You <u> must not </u> have a waterbed.

1. I <u>can</u> have a cat in my apartment. I _____ have a dog.

2. You <u>should</u> read the lease carefully. You _____ sign it without reading it.

3. The landlord <u>must</u> install a smoke detector. You _____ remove it.

4. You <u>may</u> have visitors in your apartment. You _____ make a lot of noise and disturb your neighbors.

5. If you damage something, the landlord <u>can</u> keep part of your deposit. He _____ keep all of your deposit.

6. You <u>might</u> get back all of your security deposit. If you leave your apartment in bad condition, you _____ get all of it back.

8.3 | Statements and Questions with Modals

Compare **affirmative** statements and questions with a modal.						
Wh– Word	Modal	Subject	Modal	Verb (base form)	Complement	Short Answer
		He	**can**	have	a cat in his apartment.	
	Can	he		have	a waterbed?	No, he **can't.**
What	can	he		have	in his apartment?	
		Who	**can**	have	a dog?	

Compare **negative** statements and questions with a modal.					
Wh– Word	Modal	Subject	Modal	Verb (base form)	Complement
		He	shouldn't	pay	his rent late.
Why	shouldn't	he		pay	his rent late?

EXERCISE **2**   Read each statement. Fill in the blanks to complete the question.

EXAMPLE   You should read the lease before you sign it. Why ____should I____ read the lease before I sign it?

1. You can't have a waterbed. Why _____ a waterbed?
2. We must pay a security deposit. How much _____?
3. Someone must install a smoke detector. Who _____ a smoke detector?
4. The landlord can't refuse to rent to a person because of race, religion, or nationality. Why _____ to rent to a person for these reasons?
5. Tenants shouldn't make a lot of noise in their apartments. Why _____ a lot of noise?
6. I may have a cat in my apartment. _____ have a dog in my apartment?
7. The landlord can have a key to my apartment. _____ _____ enter my apartment when I'm not home?

8.4 | *Must, Have, To, Have Got To*

Must has a very official tone. For nonofficial situations, we usually use *have to* or *have got to*.	

Examples	Explanation
The landlord **must** give you a smoke detector. The tenant **must** pay the rent on the first of each month.	For formal obligations, use *must*. *Must* is often used in legal contracts, such as apartment leases.
The landlord **has to** give you a smoke detector. The landlord **has got to** give you a smoke detector.	In conversation or informal writing, we usually use *have to* or *have got to*, not *must*.
You **must** leave the building immediately. It's on fire! You **have to** leave the building immediately. It's on fire! You**'ve got to** leave the building immediately. It's on fire!	*Must, have to,* and *have got to* express a sense of urgency. All three sentences to the left have the same meaning. *Have got to* is usually contracted: I have got to = I've got to He has got to = He's got to
Our apartment is too small. We **have to** move. Our apartment is too small. We**'ve got to** move. The landlord **has to** give you a smoke detector.	Avoid using *must* for personal obligations. it sounds very official or urgent and is too strong for most situations. Use *have to* or *have got to*. (You can use *have to* in formal situations. But don't use *must* in informal situations.)
At the end of my lease last May, I **had to** move. I **had to** find a bigger apartment.	*Must* has no past form. The past of both *must* and *have to* is *had to*. *Have got to* has no past form.

Language Notes:
1. In fast, informal speech, *have to* is often pronounced "hafta." *Has to* is often pronounced "hasta." *Got to* is often pronounced "gotta." Listen to your teacher pronounce the sentences in the above box.
2. We don't usually use *have got to* for questions and negatives.

EXERCISE **3**   Fill in the blanks with an appropriate verb. Answers may vary.

EXAMPLE   The landlord must _____give_____ you heat in cold weather.

1. You must _____ the lease with a pen. A pencil is not acceptable.
2. The landlord must _____ your security deposit if you leave your apartment in good condition.

3. The landlord must _____ you if he wants you to leave at the end of your lease.

4. You must _____ quiet in your apartment at night. Neighbors want to sleep.

5. To get a driver's license, you must _____ a driving test.

6. When you are driving, you must _____ your seat belt.

7. When you see a red light, you must _____ .

EXERCISE **4**   ABOUT YOU Make a list of personal necessities you have.

EXAMPLE   I have to change the oil in my car every three months.

1. _____

2. _____

3. _____

EXERCISE **5**   ABOUT YOU Make a list of things you had to do last weekend.

EXAMPLE   I had to do my laundry.

1. _____

2. _____

3. _____

EXERCISE **6**   Finish these statements. Practice *have got to*. Answers will vary.

EXAMPLE   When you live in the U.S., you've got to learn English.

1. When I don't know the meaning of a word, I've got to _____

_____

2. English is so important in the U.S. We've got to _____

_____

3. For this class, you've got to _____

_____

4. If you rent an apartment, you've got to _____

_____

5. If you want to drive a car, you've got to _____

_____

8.5 | Obligation with *Must* or *Be Supposed To*

Examples	Explanation
Landlord to tenant: "You **must** pay your rent on the first of each month." Judge to landlord: "You have no proof of damage. You **must** return the security deposit to your tenant."	*Must* has an official, formal tone. A person in a position of authority (like a landlord or judge) can use *must*. Legal documents use *must*.
You**'re supposed to** put your name on your mailbox. The landlord **is supposed to** give you a copy of the lease.	Avoid using *must* if you are not in a position of authority. Use *be supposed to.*
We**'re not supposed to** have cats in my building, but my neighbor has one. The landlord **was supposed to** return my security deposit, but he didn't. I**'m supposed to** pay my rent on the first of the month, but sometimes I forget.	*Be supposed to*, not *must*, is used when reporting on a law or rule that was broken or a task that wasn't completed.

Pronunciation Note:
The **d** in *supposed to* is not pronounced.

EXERCISE **7** Make these sentences less formal by changing from *must* to *be supposed to.*

EXAMPLE
You must wear your seat belt.
You're supposed to wear your seat belt.

1. You must carry your driver's license with you when you drive.
2. You must stop at a red light.
3. We must put money in the parking meter during business hours.
4. Your landlord must notify you if he wants you to leave.
5. The landlord must give me a smoke detector.
6. The teacher must give a final grade at the end of the semester.
7. We must write five compositions in this course.
8. We must bring our books to class.

EXERCISE 8 Finish these statements. Use *be supposed to* plus a verb. Answers may vary.

EXAMPLE I _'m supposed to pay my rent_____ on the first of the month.

1. Pets are not permitted in my apartment. I (not) _____
_____ a pet.

2. The landlord _____ us heat in the winter months.

3. The tenants _____ before they move out.

4. The landlord _____ a smoke detector in each apartment.

5. I _____ my rent last week, but I forgot.

6. My stove isn't working. My landlord _____ it.

7. We're going to move out next week. Our apartment is clean and in good condition. The landlord _____ our security deposit.

EXERCISE 9 ABOUT YOU Write three sentences to tell what you are supposed to do for this course. You may work with a partner.

EXAMPLE _We're supposed to write three compositions this semester._

1. _____
_____

2. _____
_____

3. _____
_____

8.6 | *Can, May, Could,* and Alternate Expressions

Example with a Modal	Alternate Expression	Explanation
I **can** clean the apartment by Friday.	It **is possible** (for me) **to** clean the apartment by Friday.	Possibility
I **can**'t understand the lease.	I **am not able to** understand the lease.	Ability
I **can**'t have a pet in my apartment.	I **am not permitted to** have a pet. I **am not allowed to** have a pet.	Permission
The landlord **may not** keep my deposit if my apartment is clean and in good condition.	The landlord **is not permitted to** keep my deposit. The landlord **is not allowed to** keep my deposit.	Permission
I **couldn**'t speak English five years ago, but I can now.	I **wasn**'t **able to** speak English five years ago, but I can now.	Past Ability
I **could** have a dog in my last apartment, but I can't have one in my present apartment.	I **was permitted to** have a dog in my last apartment, but I can't have one in my present apartment.	Past Permission

Language Notes:

1. *Can* is not usually stressed in affirmative statements. Sometimes it is hard to hear the final *t,* so we must pay attention to the vowel sound to hear the difference between *can* and *can't.* Listen to your teacher pronounce these sentences:

    I can gó. /kIn/

    I cán't go. /kænt/

    In a short answer, we pronounce *can* as /kæn/.

    Can you help me later?

    Yes, I can. /kæn/

2. We use *can* in the following common expression:

    I *can't afford* a bigger apartment. I don't have enough money.

EXERCISE **10**   Fill in the blanks with an appropriate permission word to talk about what is or isn't permitted at this school.

EXAMPLES   We ___aren't allowed to___ bring food into the classroom.

We _____can_____ leave the room without asking the teacher for permission.

1. We _____ eat in the classroom.
2. Students _____ talk during a test.
3. Students _____ use their dictionaries when they write compositions.
4. Students _____ write a test with a pencil.
5. Students _____ sit in any seat they want.
6. Students _____ use their textbooks during a test.

EXERCISE **11**   Complete each statement. Answers may vary.

EXAMPLE   The landlord may not ___refuse to rent___ to a person because of his or her nationality.

1. The tenants may not _____ the locks without the landlord's permission.
2. Each tenant in my building has a parking space. I may not _____ _____ in another tenant's space.
3. Students may not _____ during a test.
4. Teacher to students: "You don't need my permission to leave the room. You may _____ the room if you need to."
5. Some teachers do not allow cell phones in class. In Mr. Klein's class, you may not _____ during class.
6. My teacher says that after we finish a test, we may _____. We don't have to stay in class.

EXERCISE 12 ABOUT YOU Write statements to tell what is or is not permitted in this class, in the library, at this school, or during a test. If you have any questions about what is permitted, write a question for the teacher. You may work with a partner.

EXAMPLES We aren't allowed to talk in the library.

May we use our textbooks during a test?

_____

_____

_____

EXERCISE 13 ABOUT YOU Write three sentences telling about what you couldn't do in another class or school that you attended.

EXAMPLE In my high school, I couldn't call a teacher by his first name, but I can do it here.

1. _____

_____

2. _____

_____

3. _____

_____

EXERCISE 14 ABOUT YOU If you come from another country, write three sentences telling about something that was prohibited there that you can do in the U.S.

EXAMPLE I couldn't criticize the political leaders in my country, but I can do it now.

1. _____

_____

2. _____

_____

3. _____

_____

Before You Read

1. What are some complaints you have about your apartment? Do you ever tell the landlord about your complaints?

2. Is your apartment warm enough in the winter and cool enough in the summer?

 Read the following conversation. Pay special attention to *should* and *had better*.

brochure

A: My apartment is always too cold in the winter. I've got to move.

B: You don't have to move. The landlord is supposed to give enough heat.

A: But he doesn't.

B: You **should** talk to him about this problem.

A: I did already. The first time I talked to him, he just told me I **should** put on a sweater. The second time I said, "You **'d better** give me heat, or I'm going to move."

B: You **shouldn't** get so angry. That's not the way to solve the problem. You know, there are laws about heat. You **should** get information from the city so you can know your rights.

A: How can I get information?

B: Tomorrow morning, you **should** call the mayor's office and ask for a brochure about tenants' rights. When you know what the law is exactly, you **should** show the brochure to your landlord.

A: And what if he doesn't want to do anything about it?

B: Then you **should** report the problem to the mayor's office.

A: I'm afraid to do that.

B: Don't be afraid. You have rights. Maybe you **should** talk to other tenants and see if you can do this together.

8.7 | *Should; Had Better*

Examples	Explanation

You **should** talk to the landlord about the problem.
You **should** get information about tenants' rights.
You **shouldn't** get so angry.
	For advice, use *should*.
	   Should = It's a good idea.
	   Shouldn't = It's a bad idea.

Compare: Your landlord **must** give you a smoke detector. You **should** check the battery in the smoke detector occasionally.	Remember, *must* is very strong and is not for advice. It is for rules and laws. For advice, use *should*.
You **had better** give me heat, or I'm going to move. We'**d better not** make so much noise, or our neighbors will complain.	For a warning, use *had better (not)*. Something bad can happen if you don't follow this advice. The contraction for *had* (in *had better*) is '*d*. I'd you'd he'd she'd we'd they'd

Pronunciation Note:

Native speakers often don't pronounce the *had* or '*d* in *had better*. You will hear people say,
  "**You better** be careful; **You better** not make so much noise."

EXERCISE **15**   Give advice using *should*. Answers may vary.

EXAMPLE   I'm going to move next week, and I hope to get my security deposit back.

    Advice: You should clean the apartment completely.

1. I just rented an apartment, but the rent is too high for me alone.
   Advice: _____

2. My upstairs neighbors make a lot of noise.
   Advice: _____

3. The battery in the smoke detector is old.
   Advice: _____

4. I want to paint the walls.
   Advice: _____

5. The rent was due last week, but I forgot to pay it.
   Advice: _____

6. My landlady doesn't give us enough heat in the winter.
   Advice: _____

7. I can't understand my lease.

    Advice: _____

8. I broke a window in my apartment.

    Advice: _____

9. My landlord doesn't want to return my security deposit.

    Advice: _____

10. The landlord is going to raise the rent by 40 percent.

    Advice: _____

EXERCISE 16 Fill in the blanks with an appropriate verb (phrase) to complete this conversation. Answers may vary.

A: My mother is such a worrier.

B: What does she worry about?

A: Everything. Especially me.

B: For example?

A: Even if it's warm outside, she always says, "you'd

    better _____take a sweater_____

                  (example)

    because it might get cold later," or "You'd better

    _____

              (1)

    because it might rain." When I drive, she always tells

    me, "You'd better_____, or you

                  (2)

    might get a ticket." If I stay out late with my friends, she tells me,

    "You'd better _____ , or you

                  (3)

    won't get enough sleep." If I read a lot, she says, "You'd better not

    _____ , or you'll ruin your

                  (4)

    eyesight."

B: Well, she's your mother. So naturally she worries about you.

A: But she worries about other things too.

B: Like what?

A: You'd better _____ your shoes
(5)

when you enter the apartment, or the neighbors downstairs will hear
us walking around.

We'd better _____ , or the
(6)

neighbors will complain about the noise in our apartment.

B: It sounds like she's a good neighbor.

A: That's not all. She unplugs the TV every night. She says, "I'd better

_____ , or the apartments will fill
(7)

up with radiation."

And she doesn't want to use a cell phone. She says it has too much
radiation. I think that's so silly.

B: I don't think that's silly. You'd better

_____ some articles about cell
(8)

phones because they do produce radiation.

A: I don't even use my cell phone very much. But my mother always
tells me, "You'd better _____ in
(9)

case I need to call you."

B: Do you live with your mother?

A: Yes, I do. I think I'd better _____
(10)

to my own apartment, or she'll drive me crazy.

8.8 | Negatives of Modals

In negative statements, *must not, don't have to,* and *shouldn't* have very different meanings. *Must not, may not,* and *can't* have similar meanings.

Examples	Explanation
You **must not** change the locks without the landlord's permission. You **must not** take the landlord's smoke detector with you when you move.	Use *must not* for prohibition. These things are against the law or rule. *Must not* has an official tone.
I **can't** have a dog in my apartment. I **may not** have a waterbed in my apartment.	Use *cannot* or *may not* to show no permission. The meaning is about the same as *must not.* (*May not* is more formal than *cannot.*)
The landlord **is not supposed to** keep your security deposit. You **are not supposed to** paint the walls without the landlord's permission. You **are not supposed to** park in another tenant's parking space.	*Be not supposed to* indicates that something is against the law or breaks the rules. *Be not supposed to* is more common than *must not.* Remember, *must not* has an official tone.
My landlord offered me a new lease. I **don't have to** move when my lease is up. The janitor takes out the garbage. I **don't have to** take it out.	*Not have to* indicates that something is not necessary, not required. A person can do something if he wants, but he has no obligation to do it.
If you turn on the air-conditioning, you **shouldn't** leave the windows open. You **shouldn't** make noise late at night.	*Shouldn't* is for advice, not for rules.
You**'d better not** play your music so loud, or your neighbors will complain to the landlord.	*Had better not* is used for a warning.
Compare: a. I **don't have to pay** my rent with cash. I can use a check. b. I **must not pay** my rent late. c. I **don't have to use** the elevator. I can use the stairs. d. There's a fire in the building. You **must not use** the elevator.	In affirmative statements, *must* and *have to* have very similar meanings. However, in negative statements, the meaning is very different. In the examples on the left: a. It is not necessary to use cash. b. It is against the rules to pay late. c. You have options: stairs or elevator. d. It is prohibited or dangerous to use the elevator.
Compare: a. You **must not change** the locks without permission from the landlord. b. You**'re not supposed to change** the locks without permission from the landlord. c. You **shouldn't leave** your door unlocked. A robber can enter your apartment.	a. *Must not* is for prohibition. b. *Be not supposed to* is also for prohibition, but it sounds less official or formal. c. *Should not* expresses that something is a bad idea. It does not express prohibition.

EXERCISE 17 Practice using *must not* for prohibition. Use *you* in the impersonal sense.

EXAMPLE Name something you must not do.
You must not steal.

1. Name something you must not do on the bus.
2. Name something you mustn't do during a test.
3. Name something you mustn't do in the library.
4. Name something you must not do in the classroom.
5. Name something you mustn't do on an airplane.

EXERCISE 18 ABOUT YOU Tell if you *have to* or *don't have to* do the following. For affirmative statements, you can also use *have got to*.

EXAMPLES work on Saturdays
I have to work on Saturdays.   OR   I've got to work on Saturdays.

wear a suit to work
I don't have to wear a suit to work.

1. speak English every day
2. use a dictionary to read the newspaper
3. pay rent on the first of the month
4. type my homework
5. work on Saturdays
6. come to school every day
7. pay my rent in cash
8. use public transportation
9. talk to the teacher after class
10. cook every day

EXERCISE 19   Ask a student who comes from another country these questions.

1. In your native country, does a citizen have to vote?
2. Do men have to serve in the military?
3. Do schoolchildren have to wear uniforms?
4. Do divorced fathers have to support their children?
5. Do people have to get permission to travel?
6. Do students have to pass an exam to get their high school or university diploma?
7. Do students have to pay for their own books?
8. Do citizens have to pay taxes?
9. Do people have to make an appointment to see a doctor?

EXERCISE 20   Fill in the blanks with *be not supposed to* (when there is a rule) or *don't have to* (when something is not necessary).

A: Would you like to see my new apartment?

B: Yes.

A: I'll take you there after class today. The teacher says we

<u>   don't have to   </u> go to the lab this afternoon. We
      (example)

can take the day off today.

(at the apartment)

B: Why do you carry your bicycle up to the third floor? Wouldn't it be better to leave it near the front door?

A: The landlord says we _____ leave anything near
                                 (1)

the door.

The rule is to leave the front lobby empty. Besides, I can take it up in

the elevator. I _____ use the stairs, but I don't
                         (2)

mind carrying it. My bicycle is light.

B: This is a great apartment. But it's so big. Isn't it expensive?

A: Yes, but I _____ pay the rent alone. I have a
                     (3)

roommate.

B: I see you have lots of nice pictures on the walls. In my apartment,

we _____ make holes in the walls.
              (4)

A: You can't even put up pictures? If you use picture hooks, you _____
_____ make big holes. Why don't you ask your land-
(5)

lord if you can do it? If you can, I can give you some picture hooks.

B: Thanks. In my apartment, the landlord has so many rules. For
example, we _____ hang our laundry out the
(6)

window. We have to use the washing machine in the basement. And
we _____ use electric heaters.
(7)

A: An electric heater can sometimes cause a fire. I'm sure the apartment
has heaters for each room. And in the U.S. people don't usually
hang clothes to dry out the window. People use driers.

B: There are so many different rules and customs here.

A: Don't worry. If you do something wrong, someone will tell you.

EXERCISE **21**

Students (S) are asking the teacher (T) questions about the final
exam. Fill in the blanks with the negative form of *have to*, *should*,
must, had better, can, may, be supposed to. In some cases, more
than one answer is possible.

S: Do I have to sit in a specific seat for the test?

T: No, you ___ don't have to ___. You can choose any seat you want.
(example)

S: Is it OK if I talk to another student during a test?

T: No. Absolutely not. You _____ talk to another student
(1)

during a test.

S: Is it OK if I use my book?

T: Sorry. You _____ use your book.
(2)

S: What if I don't understand something on the test? Can I ask another
student?

T: You _____ talk to another student, or I'll think you're
(3)

getting an answer. Ask me if you have a question.

S: What happens if I am late for the test? Will you let me in?

T: Of course I'll let you in. But you _____ come late.
    (4)

    You'll need a lot of time for the test.

S: Do I have to bring my own paper for the final test?

T: If you want to, you can. But you _____ bring paper.
    (5)

    I'll give you paper if you need it.

S: Must I use a pen?

T: You can use whatever you want. You _____ use
    (6)

    a pen.

S: Do you have any advice on test-taking?

T: Yes. If you see an item that is difficult for you, go on to the next

    item. You _____ spend too much time on a
    (7)

    difficult item, or you won't finish the test.

S: Can I bring coffee into the classroom?

T: The school has a rule about eating or drinking in the classroom. You

    _____ bring food into the classroom.
    (8)

S: If I finish the test early, must I stay in the room?

T: No, you _____ stay. You can leave.
    (9)

THE NEW NEIGHBORS

Before You Read

1. Are people friendly with their neighbors in your community?

2. Do you know any of your neighbors now?

Read the following conversation. Pay special attention to *must*.

crib

Lisa (L) knocks on the door of her new upstairs neighbor, Paula (P).

L: Hi. You **must be** the new neighbor. I saw the moving truck out front this morning. Let me introduce myself. My name is Lisa. I live downstairs from you.

P: Nice to meet you, Lisa. My name is Paula. We just moved in.

L: I saw the movers carrying a crib upstairs. You **must have** a baby.

P: We do. We have a ten-month-old son. He's sleeping now. Do you have any kids?

L: Yes. I have a 16-year-old-daughter and an 18-year-old son.

P: It **must be** hard to raise teenagers.

L: Believe me, it is! I **must spend** half my time worrying about where they are and what they're doing. My daughter talks on the phone all day. She **must spend** half of her waking hours on the phone with her friends. They're always, whispering to each other. They **must have** some big secrets.

P: I know what you mean. My brother has a teenage daughter.

L: Listen, I don't want to take up any more of your time. You **must be** very busy. I just wanted to bring you these cookies.

P: Thanks. That's very nice of you. They're still warm. They **must be** right out of the oven.

L: They are. Maybe we can talk some other time when you're all unpacked.

8.9 | *Must* for Conclusions

In Section 8.4, we studied *must* to express necessity. *Must* has another use: we use it to show a logical conclusion or deduction based on information we have or observations we make. *Must*, in this case, is for the present only, not the future.

Examples	Explanation
a. The new neighbors have a crib. They **must have** a baby. b. Paula just moved in. She **must be** very busy. c. The teenage girls whisper all the time. They **must have** secrets.	a. You see the crib, so you conclude that they have a baby. b. You know how hard it is to move, so you conclude that she is busy. c. You see them whispering, so you conclude that they are telling secrets.
I didn't see Paula's husband. He **must not** be home.	For a negative deduction, use *must not*. Do not use a contraction.

EXERCISE 22 A week later, Paula goes to Lisa's apartment and notices certain things. Use *must* + base form to show Paula's conclusions about Lisa's life. Answers may vary.

EXAMPLE There is a bowl of food on the kitchen floor.

Lisa must have a pet.

1. There are pictures of Lisa and her two children all over the house. There is no picture of a man.

   _____

2. There is a nursing certificate on the wall with Lisa's name on it.

   _____

3. There are many different kinds of coffee on a kitchen shelf.

   _____

4. There are a lot of classical music CDs.

   _____

5. In Lisa's bedroom, there's a sewing machine.

   _____

6. In the kitchen, there are a lot of cookbooks.

   _____

Family Calendar			
Week 1	Mom	Joey	Ann
Sunday	Church social		dinner
Monday	Carpool to school		Piano lesson
Tuesday	PTA Meeting	Art Class	
Wednesday	Take cat to Vet	Start Science project	Report Due
Thursday	Carpool to school		Library with Lila
Friday		Dentist	Dentist
Saturday		School play rehearsal	

7. There's a piano in the living room.

_____

8. On the bookshelf, there are a lot of books about modern art.

_____

9. On the kitchen calendar, there's an activity filled in for almost every day of the week.

_____

10. There are pictures of cats everywhere.

_____

EXERCISE 23

Two neighbors, Alma (A) and Eva (E), meet in the hallway of their building. Fill in the blanks with an appropriate verb to show deduction.

A: Hi. My name's Alma. I live on the third floor. You must

_____be_____ new in this building.
(example)

E: I am. We just moved in last week. My name's Eva.

A: I noticed your last name on the mailbox. It's Ković. That sounds like a Bosnian name. You must _____ from Bosnia.
(1)

E: I am. How did you know?

A: I'm from Bosnia too. Did you come directly to the U.S. from Bosnia?

E: No. I stayed in Germany for three years.

A: Then you must _____ German.
(2)

E: I can speak it pretty well, but I can't write it well.

A: Are you going to school now?

E: Yes, I'm taking English classes at Washington College.

A: What level are you in?

E: I'm in Level 5.

A: Then you must _____ my husband. He takes classes
(3)

there too. He's in Level 5 too.

E: There's only one guy with a Bosnian last name. That must _____

_____ your husband.

(4)

A: His name is Hasan.

B: Oh, yes, I know him. I didn't know he lived in the same building. I never see him here. He must not _____ home very much.

(5)

A: He isn't. He has two jobs.

E: Do you take English classes?

A: Not anymore. I came here 15 years ago.

E: Then your English must _____ perfect.

(6)

A: I don't know if it's perfect, but it's good enough.

8.10 | *Will* and *May / Might*

Examples	Explanation
My lease **will** expire on April 30. We **won't** sign another lease.	For certainty about the future, use *will*. The negative contraction for *will not* is *won't*.
a. My landlord **might** raise my rent at that time. a. I **may** move. b. I don't know what "tenant" means. Let's ask the teacher. She **might** know. b. The teacher **may** have information about tenants' rights.	*May* and *might* both have about the same meaning: possibility or uncertainty. a. about the future b. about the present
He **may not** renew our lease. He **might not** renew our lease.	We don't use a contraction for *may not* and *might not*.

Compare:
a. **Maybe** I will move.
b. I **might** move.
a. **Maybe** he doesn't understand the lease.
b. He **might** not understand the lease.
a. **Maybe** the apartment is cold in winter.
  (*maybe* = adverb)
b. The apartment **may** be cold in winter.
  (*may + be* = modal + verb)
	Maybe is an adverb. It is one word. It usually comes at the beginning of the sentence and means *possibly* or *perhaps*. *May* and *might* are modals. They follow the subject and precede the verb. Sentences (a) and (b) have the same meaning.
	  Wrong: I *maybe* will move.
	  Wrong: He *maybe* doesn't understand.
	  Wrong: The apartment *maybe* is cold.

EXERCISE 24   The following sentences contain *maybe*. Take away *maybe* and use *may* or *might* + base form.

EXAMPLE   Maybe your neighbors will complain if your music is loud.
Your neighbors might complain if your music is loud.

1. Maybe my sister will come to live with me.
2. Maybe she will find a job in this city.
3. Maybe my landlord will raise my rent.
4. Maybe I will get a dog.
5. Maybe my landlord won't allow me to have a dog.
6. Maybe I will move next year.
7. Maybe I will buy a house soon.
8. Maybe I won't stay in this city.
9. Maybe I won't come to class tomorrow.
10. Maybe the teacher will review modals if we need more help.

EXERCISE 25   ABOUT YOU Fill in the blanks with a possibility.

EXAMPLES   If I don't pay my rent on time, <u>I might have to pay a late fee.</u>
If I make a lot of noise in my apartment, <u>the neighbors may complain.</u>

1. When my lease is up, _____
2. If I don't clean my apartment before I move out, _____

   _____
3. If I don't study for the next test, _____
4. If we don't register for classes early, _____

   _____
5. If I don't pass this course, _____

EXERCISE 26   Fill in the blanks with possibilities. Answers may vary.

EXAMPLE   A: I'm going to move on Saturday. I might _____need_____ help. Can you help me?

B: I'm not sure. I may _____go_____ to the country with my family if the weather is nice. If I stay here, I'll help you.

1. A: My next door neighbor's name is Terry Karson. I see her name on the doorbell but I never see her.

   B: Why do you say "her"? Your neighbor may _____.
   Terry is sometimes a man's name.

2. A: I need coins for the laundry room. Do you have any?

   B: Let me look. I might _____ some. No, I don't have any. Look in the laundry room. There might _____ a dollar-bill changer there.

3. A: Do you know the landlord's address?

   B: No, I don't. Ask the manager. She might _____.

   A: Where's the manager now?

   B: I'm not sure. She might _____ in a tenant's apartment.

4. A: Do they allow cats in this building?

   B: I'm not sure. I know they don't allow dogs, but they might _____ cats.

5. A: We'd better close the windows before going out.

   B: Why? It's a hot day today.

   A: Look how gray the sky is. It might _____.

6. A: Are you going to stay in this apartment for another year?

   B: I'm not sure. I may _____.

   A: Why?

   B: The landlord might _____ the rent. If the rent goes up more than 25 percent, I'll move.

7. A: I have so much stuff in my closet. There's not enough room for my clothes.

   B: There might _____ lockers in the basement where you can store your things.

   A: Really? I didn't know that.

   B: Let's look. I may _____ a key to the basement with me.

   A: That would be great.

   B: Hmm. I don't have one on me. Let's go to my apartment. My basement keys might _____ there.

Before You Read

1. People often have a garage sale or yard sale or an apartment sale before they move. At this kind of sale, people sell things that they don't want or need anymore. Did you ever buy anything at this kind of sale?

2. At a garage or yard sale, it is usually not necessary to pay the asking price. You may be able to bargain[3] with the seller. Can you bargain the price in other places?

This is a conversation at a garage sale between a seller (S) and a buyer (B). Read the conversation. Pay special attention to modals and related expressions.

S: I see you're looking at my microwave oven. **May** I answer any questions?

B: Yes. I'm interested in buying one. Does it work well?

S: It's only two years old, and it's in perfect working condition. **Would** you **like** to try it out?

B: Sure. **Could** you plug it in somewhere?

S: I have an outlet right here. **Why don't we** boil a cup of water so you can see how well it works.

A few minutes later...

B: It seems to work well. **Would** you tell me why you're selling it, then?

S: We're moving next week. Our new apartment already has one.

B: How much do you want for it?[4]

S: $40.

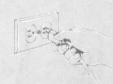

outlet

---

[3] When a buyer *bargains* with the seller, the buyer makes an offer lower than the asking price and hopes that he or she and the seller will agree on a lower price.

[4] We ask "How much is it?" when the price is fixed. We ask "How much do you want for it?" when the price is negotiable—you can bargain for it.

B: **Will** you take $30?

S: **Can** you wait a minute? I'll ask my wife.

A few minutes later...

S: My wife says she'll let you have it for $35.

B: OK. **May** I write you a check?

S: I'm sorry. I**'d rather** have cash.

B: **Would** you hold it for me for an hour? I can go to the ATM and get cash.

S: **Could** you leave me a small deposit? Ten dollars, maybe?

B: Yes, I can.

S: Fine. I'll hold it for you.

8.11 | Using Modals and Questions for Politeness

Modals and questions are often used to make direct statements more polite.
Compare:

    Plug it in. (very direct)

    Would you plug it in? (more polite)

	Examples	Explanation
To ask permission	**May** **Can** **Could** } I write you a check?	*May* and *could* are considered more polite than *can* by some speakers of English.
To request that someone do something	**Can** **Could** **Will** **Would** } you plug it in?	For a request, *could* and *would* are softer than *can* and *will*.
To express want or desire	**Would** you **like** to try out the microwave oven? Yes, I **would like** to see if it works. I**'d like** a cup of coffee.	*Would like* has the same meaning as *want*. *Would like* is softer than *want*. The contraction for *would* after a pronoun is *'d.*
To express preference	**Would** you **rather** pay with cash or by credit card? I**'d rather** pay by credit card (than with cash).	Use *or* in questions with *would rather*. Use *than* in statements.
To offer a suggestion	**Why don't you** go to the ATM to get cash? **Why don't we** boil a cup of water? Compare: Go to the ATM. Boil a cup of water.	We can make a suggestion more polite by using a negative question.

EXERCISE **27**   Change each request to make it more polite.   Practice *may,   can,* and *could* + *I*?

  EXAMPLES   I want to use your phone.
             May I use your phone?

             I want to borrow a quarter.
             Could I borrow a quarter?

             1. I want to help you.   3. I want to leave the room.
             2. I want to close the door.   4. I want to write you a check.

EXERCISE **28**   Change these commands to make them more polite.   Practice *can you, could you, will you,* and *would you?*

  EXAMPLE   Call the doctor for me.
            Would you call the doctor for me?

            Give me a cup of coffee.
            Could you give me a cup of coffee, please?

            1. Repeat the sentence.   3. Spell your name.
            2. Give me your paper.   4. Tell me your phone number

EXERCISE **29**   Make these sentences more polite by using *would like*.

  EXAMPLE   Do you want some help?
            Would you like some help?

            1. I want to ask you a question.
            2. The teacher wants to speak with you.
            3. Do you want to try out the oven?
            4. Yes. I want to see if it works.

EXERCISE **30**   Make each suggestion more polite by putting it in the form of a negative question.

  EXAMPLES   Plug it in.
             Why don't you plug it in?

             Let's eat now.
             Why don't we eat now?

             1. Take a sweater.   3. Turn left here.
             2. Let's turn off the light. 4. Let's leave early.

EXERCISE **31**   ABOUT YOU Make a statement of preference using *would rather*.

EXAMPLE   own a house / a condominium
I'd rather own a condominium (than a house).

1. live in the U.S. / in another country
2. own a condominium / rent an apartment
3. have young neighbors / old neighbors
4. have wood floors / carpeted floors
5. live in the center of the city / in a suburb
6. drive to work / take public transportation
7. pay my rent by check / cash
8. have nosy neighbors / noisy neighbors

EXERCISE **32**   ABOUT YOU Ask a question of preference with the words given. Another student will answer.

EXAMPLE   eat Chinese food / Italian food
A: Would you rather eat Chinese food or Italian food?
B: I'd rather eat Italian food.

1. read fact / fiction
2. watch funny movies / serious movies
3. listen to classical music / popular music
4. visit Europe / Africa
5. own a large luxury car / a small sports car
6. watch a soccer game / take part in a soccer game
7. write a letter / receive a letter
8. cook / eat in a restaurant

EXERCISE 33   This is a conversation between a seller (S) and a buyer (B) at a garage sale. Make this conversation more polite by using modals and other polite expressions in place of the underlined words. Answers may vary.

S: *May I help you?*
~~What do you want?~~
(example)

B: I'm interested in that lamp. <u>Show it to me</u>. Does it work?
                         (1)

S: I'll go and get a light bulb. <u>Wait a minute</u>.
                              (2)

A few minutes later...

B: <u>Plug it in</u>.
    (3)

S: You see? It works fine.

B: How much do you want for it?

S: This is one of a pair. I have another one just like it. They're $10 each. I <u>prefer to sell</u> them together.
            (4)

B: <u>Give them both to me for $15</u>.
                (5)

S: I'll have to ask my husband.

    (A few seconds later)

My husband says he'll sell them to you for $ 17.

B: Fine. I'll take them. Will you take a check?

S: I <u>prefer to</u> have cash.
      (6)

B: I only have five dollars on me.

S: OK. I'll take a check. <u>Show me some identification</u>.
                        (7)

B: Here's my driver's license.

S: That's fine. Write the check to James Kucinski.

B: <u>Spell your name for me</u>.
          (8)

S: K-U-C-I-N-S-K-I.

Modals		
Modal	Example	Explanation
can	I **can** stay in this apartment until March.	Permission
	I **can** carry my bicycle up to my apartment.	Ability/Possibility
	You **can**'t paint the walls without the landlord's permission.	Prohibition
	Can I borrow your pen?	Asking permission
	Can you turn off the light, please?	Request
should	You **should** be friendly with your neighbors.	A good idea
	You **shouldn**'t leave the air-conditioner on. It wastes electricity.	A bad idea
may	**May** I borrow your pen?	Asking permission
	You **may** leave the room.	Giving permission
	You **may not** talk during a test.	Prohibition
	I **may** move next month.	Future possibility
	The landlord **may** have an extra key.	Present possibility
might	I **might** move next month.	Future possibility
	The landlord **might** have an extra Key.	Present possibility
must	The landlord **must** install smoke detectors.	Rule or law: Official tone
	You **must not** change the locks.	Prohibition: Official tone
	Mary has a cat box. She **must** have a cat.	Conclusion/Deduction
would	**Would** you help me move?	Request
would like	I **would like** to use your phone.	Want
would rather	I **would rather** live in Florida than in Maine.	Preference
could	In my country, I **couldn**'t choose my own apartment. The government gave me one.	Past permission
	In my country, I **could** attend college for free.	Past ability
	Could you help me move?	Request
	Could I borrow your car?	Asking permission

Related Expressions		
Expression	Example	Explanation
have to	She **has to** leave. He **had to** leave work early today.	Necessity Past necessity
have got to	She **has got to** see a doctor. I'**ve got to** move.	Necessity
not have to	You **don't have to** pay your rent with cash. You can pay by check.	No necessity
had better	You **had better** pay your rent on time, or the landlord will ask you to leave. You'**d better** get permission before changing the locks.	Warning
be supposed to	I **am supposed to** pay my rent by the fifth of the month. We'**re not supposed to** have a dog here.	Reporting a rule
be able to	The teacher **is able to** use modals correctly.	Ability
be permitted to be allowed to	We'**re not permitted to** park here overnight. We'**re not allowed to** park here overnight.	Permission

EDITING ADVICE

1. After a modal, we use the base form.

    I must ~~to~~ study.

    I can help~~ing~~ you now.

2. A modal has no ending.

    He can~~s~~ cook.

3. We don't put two modals together. We change the second modal to another form.

    She will ~~must~~ ^{have to} take the test.

4. Don't forget *to* after *be permitted, be allowed, be supposed,* and *be able.*

    We're not permitted ^to talk during a test.

5. Don't forget *be* before *permitted to, allowed to, supposed to,* and *able to.*

>   am
> I_∧not supposed to pay my rent late.

6. Use the correct word order in a question.

>   should I
> What I̶ ̶s̶h̶o̶u̶l̶d̶ do about my problem?

7. Don't use *can* for past. Use *could* + a base form.

>   couldn't go
> I c̶a̶n̶'̶t̶ ̶w̶e̶n̶t̶ to the party last week.

8. Don't forget *would* before *rather.*

>   'd
> I_∧rather live in Canada than in the U.S.

9. Don't forget *had* before *better.*

>   'd
> You_∧better take a sweater. It's going to get cold.

10. Don't forget *have* before *got to.*

>   've
> It's late. I_∧got to go.

11. Don't use *maybe* before a verb.

>   may
> It m̶a̶y̶b̶e̶ ̶w̶i̶l̶l̶ rain later.

PART **1** Find the grammar mistakes with the underlined words, and correct them. Not every sentence has a mistake. If the sentence is correct, write *C*.

EXAMPLES You must ~~to~~ stop at a red light.

You have to stop at a red light. C

1. We're not permitted use our books during the test.
2. When I was a child, I couldn't rode a bike.
3. When she can leave?
4. What must I write on this application?
5. She has to taking her daughter to the doctor now.
6. What we should do for homework?
7. You not supposed to talk during the test.
8. We're not allowed to take food into the computer lab.
9. He can't have a dog in his apartment.
10. Could I use your pen, please?
11. I rather walk than drive.
12. You'd better hurry. It's late.
13. I got to talk to my boss about a raise.
14. We maybe will buy a house.
15. She might buy a new car next year.
16. I may have to go home early tonight.
17. He can speak English now, but he can't spoke it five years ago.

PART **2** This is a conversation between two friends. Circle the correct expression in parentheses ( ) to complete the conversation.

A: I'm moving on Saturday. (*Could* / *May*) you help me?
　　　　　　　　　　　　　　(*example*)

B: I (*should* / *would*) like to help you, but I have a bad back. I went to
　　(*1*)

my doctor last week, and she told me that I (*shouldn't* / *don't have to*)
　　　　　　　　　　　　　　　　　　　　　　(*2*)

lift anything heavy for a while. (*Can / Would*) I help you any other
(3)

way besides moving?

A: Yes. I don't have enough boxes. (*Should / Would*) you help me find
(4)

some?

B: Sure. I (*have to / must*) go shopping this afternoon. I'll pick up some
(5)

boxes while I'm at the supermarket.

A: Boxes can be heavy. You (*would / had*) better not lift them yourself.
(6)

B: Don't worry. I'll have someone put them in my car for me.

A: Thanks. I don't have a free minute. I (*couldn't go / can't went*) to
(7)

class all last week. There's so much to do.

B: I know what you mean. You (*might / must*) be tired.
(8)

A: I am. I have another favor to ask. (*Can / Would*) I borrow your van
(9)

on Saturday?

B: I (*should / have to*) work on Saturday. How about Sunday? I
(10)

(*must not / don't have to*) work on Sunday.
(11)

A: That's impossible. I (*'ve got to / should*) move out on Saturday. The
(12)

new tenants are moving in Sunday morning.

B: Let me ask my brother. He has a van too. He (*must / might*) be able
(13)

to let you use his van. He (*has to / should*) work Saturday too, but
(14)

only for half a day.

A: Thanks. I'd appreciate it if you could ask him.

B: Why are you moving? You have a great apartment.

A: We decided to move to the suburbs. It's quieter there. And I want to
have a dog. I (*shouldn't / 'm not supposed to*) have a dog in my
(15)

present apartment. But my new landlord says I (*might / may*) have
(16)

a dog.

B: I (*had / would*) rather have a cat. They're easier to take care of.
(17)

Classroom Activities

1. A student will read one of the following problems out loud to the class, pretending that this is his or her problem. Other students will ask for more information and give advice about this problem.

EXAMPLE   My mother-in-law comes to visit all the time. When she's here, she always criticizes everything we do. I told my wife that I don't want her here, but she says, "It's my mother, and I want her here." What should I do?

A: How long do you think she will stay?

B: She might stay for about two weeks or longer.

C: How does she criticize you? What does she say?

B: She says I should help my wife more.

D: Well, I agree with her. You should help with house-work.

B: My children aren't allowed to watch TV after 8 o'clock. But my mother-in-law lets them watch TV as long as they want.

E: You'd better have a talk with her and tell her your rules.

Problem 1. My mother is 80 years old, and she lives with us. It's very hard on my family to take care of her. We'd like to put her in a nursing home, where she can get better care. Mother refuses to go. What can we do?

Problem 2. I have a nice one-bedroom apartment with a beautiful view of a park and a lake. I live with my wife and one child. My friends from out of town often come to visit and want to stay at my apartment. In the last year, ten people came to visit us. I like to have visitors, but sometimes they stay for weeks. It's hard on my family with such a small apartment. What should I tell my friends when they want to visit?

Problem 3. My upstairs neighbors make noise all the time. I can't sleep at night. I asked them three times to be quieter, and each time they said they would. But the noise still continues. What should I do?

Write your own problem to present to the class. It can be real or imaginary. (Suggestions: a problem with a neighbor, your landlord, a teacher or class, a service you are dissatisfied with)

_____

_____

_____

2. Circle a game you like from the following list. Find a partner who also likes this game. Write a list of some of the rules of this game. Tell what you *can, cannot, should, have to,* and *must not* do.

| chess | tennis | football | poker | other _____ |
| checkers | baseball | soccer | volleyball | |

EXAMPLE checkers

You have to move the pieces on a diagonal. You can only move in one direction until you get a king. Then you can move in two directions.

_____
_____
_____

3. Many people get vanity license plates that tell something about their professions, hobbies, or families. Often words are abbreviated: M = am, U = you, 4 = for, 8 = the "ate" sound. Words are often missing vowels. If you see the following license plates, what conclusion can you make about the owner? You may work with a partner and get help from the teacher.

EXAMPLE EYE DOC The owner must be an eye doctor. _____

1. I TCH ENGLSH _____
2. I LV CARS _____
3. I M GRANDMA _____
4. MUSC LVR _____
5. I LV DGS _____
6. TENNIS GR8 _____
7. I SK8 _____
8. CRPNTR _____
9. BSY MOM _____
10. SHY GUY _____
11. DAD OF TWO _____
12. RMNTIC GAL _____
13. NO TIME 4 U _____
14. CITY GAL _____
15. LDY DOC _____
16. LUV GLF _____
17. MXCAN GUY _____
18. I M GD COOK _____

19. ALWAYS L8 _____

20. WE DANCE _____

4. Work with a partner from your own country, if possible. Talk about some laws in your country that are different from laws in the United States. Present this information to the class.

> EXAMPLE Citizens must vote in my country. In the U.S., they don't have to vote.
> People are supposed to carry identification papers at all times. In the U.S., people don't have to carry identification papers.
> In my country, citizens must not own a gun.

Talk About it

1. Compare getting a driver's license here with getting a driver's license in another country or state. Are the requirements the same?

2. How did you find your apartment?

Write About it

1. Write a short composition comparing rules in an apartment in this city with rules in an apartment in your hometown or native country.

2. Write about the differences between rules at this school and rules at another school you attended. Are students allowed to do things here that they can't do in another school?

3. Find out what a student has to do to register for the first time at this school. You may want to visit the registrar's office to interview a worker there. Write a short composition explaining to a new student the steps for admission and registration.

Outside Activities

1. Look at the Sunday newspaper for notices about garage sales or apartment sales.   What kind of items are going to be sold?   If you have time, go to a sale. Report about your experience to the class.

2. Get a newspaper. Look for the advice column. Read the problems and the advice. Circle the modals. Do you agree with the advice?

3. Look at your lease. Can you understand what the rules are in your apartment?

Internet Activities

1. Try to find information online about tenants' rights in the city where you live. Circle the modals.

2. Find a phone directory online. Look up the names and addresses of moving companies in your city. Call a company to find out the price of a move.

3. Find apartments for rent online. Print a page. Discuss with your classmates the price of apartments and what is included.

Additional Activities at http://elt.thomson.com/gic

GRAMMAR

The Present Perfect
The Present Perfect Continuous[1]

CONTEXT: Searching the Web

Google
Genealogy

[1] The *present perfect continuous* is sometimes called the *present perfect progressive*.

9.1 | The Present Perfect Tense—An Overview

We form the present perfect with *have* or *has* + the past participle.

Subject	*have*	Past Participle	Complement	Explanation
I	have	been	in the U.S. for three years.	Use *have* with *I, you, we, they,* and plural nouns.
You	have	used	your computer a lot.	
We	have	written	a job résumé.	
They	have	bought	a new computer.	
Computers	have	changed	the world.	

Subject	*has*	Past Participle	Complement	Explanation
My sister	has	gotten	her degree.	Use *has* with *he, she, it,* and singular nouns.
She	has	found	a job as a programmer.	
My father	has	helped	me.	
It	has	rained	a lot this month.	

There	*has/have*	*been*	Complement	Explanation
There	has	been	a problem with my computer.	After *there,* we use *has* or *have*, depending on the noun that follows. Use *has* with a singular noun. Use *have* with a plural noun.
There	have	been	many changes in the world.	

GOOGLE

Before You Read

1. Do you surf the Internet a lot? Why?

2. What search engine do you usually use?

Larry Page and Sergey Brin

Read the following article. Pay special attention to the present perfect tense.

Since its start in 1998, Google **has become** one of the most popular search engines. It **has grown** from a research project in the dormitory room of two college students to a business that now employs approximately 1,000 people.

Google's founders, Larry Page and Sergey Brin, met in 1995 when they were in their 20s and graduate students in computer science at Stanford University in California. They realized that Internet search was a very important field and began working together to make searching easier. Both Page and Brin left their studies at Stanford to work on their project. Interestingly, they **have** never **returned** to finish their degrees.

Brin was born in Russia, but he **has lived** in the U.S. since he was five years old. His father was a mathematician in Russia. Page, whose parents were computer experts, **has been** interested in computers since he was six years old.

When Google started in 1998, it did 10,000 searches a day. Today it does 200 million searches a day in 90 languages. It indexes[2] three billion Web pages.

How is Google different from other search engines? **Have** you ever **noticed** how many ads and banners there are on other search engines? News, sports scores, stock prices, links for shopping, mortgage rates, and more fill other search engines. Brin and Page wanted a clean home page. They believed that people come to the Internet to search for specific information, not to be hit with a lot of unwanted data. The success of Google over its rivals[3] **has proved** that this is true.

Over the past few years, Google **has added** new features to its Web site: Google Images, where you can type in a word and get thousands of pictures; Google News, which takes you to today's news; Froogle, which takes you to a shopping site; and more. But one thing **hasn't changed**: the clean opening page that Google offers its users.

In 2003, *Fortune* magazine ranked Page and Brin among the top ten richest people under 30, So far these two men **haven't changed** their lifestyles very much. They continue to live modestly.

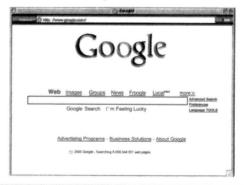

---

[2] *To index* means to sort, organize, and categorize information.

[3] *Rivals* are competitors.

EXERCISE 1 Underline the present perfect tense in each sentence. Then tell if the sentence is true or false.

EXAMPLE Google <u>has become</u> the number one search engine. T

1. Google has grown over the years.
2. Sergey Brin has lived in the U.S. all his life.
3. Larry Page and Sergey Brin have known each other since they were children.
4. Larry Page has been interested in computers since he was a child.
5. Brin and Page have returned to college to finish their degrees.
6. Since they became rich, Brin and Page have changed their lifestyles.
7. The word "Google" has become a verb.

9.2 | The Past Participle

Forms			Explanation
Regular Verbs			The past participle of regular verbs ends in -ed. The past form and the past participle for regular verbs are the same.
Base Form	Past Form	Past Participle	
work	worked	**worked**	
improve	improved	**improved**	
Irregular Verbs			The past participle of irregular verbs is sometimes the same as the past form and sometimes different from it.
Base Form	Past Form	Past Participle	For an alphabetical list of irregular past tenses and past participles, see Appendix M.
have	had	**had** (same as past)	
write	wrote	**written** (different from past)	

9.3 | Irregular Past Participle Forms of Verbs[4]

Base Form	Past Form	Past Participle
become	became	become
come	came	come
run	ran	run
blow	blew	blown
draw	drew	drawn
fly	flew	flown
grow	grew	grown
know	knew	known
throw	threw	thrown
swear	swore	sworn
tear	tore	torn
wear	wore	worn
break	broke	broken
choose	chose	chosen
freeze	froze	frozen
speak	spoke	spoken
steal	stole	stolen
begin	began	begun
drink	drank	drunk
ring	rang	rung
sing	sang	sung
sink	sank	sunk
swim	swam	swum
arise	arose	arisen
bite	bit	bitten
drive	drove	driven
ride	rode	ridden
rise	rose	risen
write	wrote	written
be	was/were	been
eat	ate	eaten
fall	fell	fallen
for give	forgave	for given
give	gave	given
mistake	mistook	mistaken
see	saw	seen
shake	shook	shaken
take	took	taken
do	did	done
forget	forgot	forgotten
get	got	gotten
go	went	gone
lie	lay	lain
prove	proved	proven (or proved)
show	showed	shown (or showed)

[4] For an alphabetical listing of irregular past tenses and past participles, see Appendix M.

EXERCISE **2**   Write the past participle of these verbs.

EXAMPLE   **eat** _____eaten_____

1. go _____
2. see _____
3. look _____
4. study _____
5. bring _____
6. take _____
7. say _____
8. be _____
9. find _____
10. leave _____
11. live _____
12. know _____
13. like _____
14. fall_____
15. feel_____

16. come _____
17. break _____
18. wear _____
19. choose _____
20. drive _____
21. write _____
22. put _____
23. begin _____
24. want _____
25. get _____
26. fly _____
27. sit _____
28. drink _____
29. grow _____
30. give _____

9.4 | The Present Perfect—Contractions, Negatives

For an affirmative statement, we can make a contraction with the subject pronoun.
For a negative statement, we can make a contraction using *have/has* + *n't*.

Examples	Explanation
I've had a lot of experience with computers. **We've** read the story about Google. **He's** been interested in computers since he was a child. **There's** been an increase in searching over the years.	We can make a contraction with subject pronouns and *have* or *has*. I have = I've He has = He's You have = You've She has = She's We have = We've It has = It's They have = They've There has = There's
Larry's lived in the U.S. all his life. **Sergey's** been in the U.S. since he was five years old.	Most singular nouns can contract with *has*.
I **haven't** studied programming. Brin **hasn't** finished his degree.	Negative contractions: *have not = haven't* *has not = hasn't*

Language Note:
 The **'s** in *he's*, *she's*, *it's*, and *there's* can mean *has* or *is*. The word following the contraction will tell you what the contraction means.
    He's working. = He is working.
    He's worked. = He *has* worked.	

EXERCISE **3**   Fill in the blanks to form the present perfect. Make a contraction, if possible.

EXAMPLE   You _'ve_____ bought a new computer.

1. I _____ learned a lot about computers.
2. We _____ read the story about Google.
3. Larry _____ known Sergey since they were at Stanford University.
4. They (not) _____ known each other since they were children.
5. It _____ been easy for me to learn about computers.
6. You _____ used the Internet many times.
7. Larry and Sergey (not) _____ finished their degrees.

9.5 | Adding an Adverb

Subject	has/ have	Adverb	Past Participle	Complement	Explanation
Page and Brin	**have**	never	**finished**	their degrees.	You can put an adverb between the auxiliary verb (*have/has*) and the past participle.
They	**have**	already*	**made**	a lot of money.	
They	**have**	even	**become**	billionaires in their 30s.	
Larry Page	**has**	always	**been**	interested in computers.	
You	**have**	probably	**used**	a search engine.	

*Note:
Already frequently comes at the end of the verb phrase.
  They have made a lot of money **already**.

EXERCISE **4**   Add the word in parentheses ( ) to the sentence.

EXAMPLE   You have gotten an e-mail account. (probably)
    You have probably gotten an e-mail account.

1. The teacher has given a test on this lesson. (not)

_____

2. We have heard of Page and Brin. (never)

_____

3. They have been interested in search technology. (always)

_____

4. You have used Google. (probably)

_____

5. Brin hasn't finished his degree. (even)

_____

9.6 | **The Present Perfect—Statements and Questions**

Compare affirmative statements and questions

Wh-Word	have/has	Subject	have/has	Past Participle	Complement	Short Answer
		Larry	has	lived	in the U.S. all his life.	
	Has	Sergey		lived	in the U.S. all his life?	No, he hasn't.
How long	has	Sergey		lived	in the U.S.?	Since 1979.

Language Note:

For a short *yes* answer, we cannot make a contraction.

Has Larry lived in the U.S. all his life? Yes, he has. (Not: *he's*)

Compare negative statements and questions

Wh-Word	haven't/hasn't	Subject	haven't/hasn't	Past Participle	Complement
		They	haven't	finished	their degrees.
Why	haven't	they		finished	their degrees?

EXERCISE **5**   Change the statement to a question.

EXAMPLE   Google has changed the way people search. (how)
How has Google changed the way people search?

1. I have used several search engines. (which ones)

_____

2. Larry and Sergey haven't finished their degrees. (why)

_____

3. They haven't changed their lifestyles. (why)

_____

4. Sergey has been in the U.S. for many years. (how long)

_____

5. Larry and Sergey have hired approximately 1,000 people to work for Google. (why)

_____

6. We have used the computer lab several times this semester. (how many times)

_____

7. The memory and speed of computers has increased. (why)

_____

8. Computers have become part of our daily lives. (how)

_____

9.7 | Continuation from Past to Present

We use the present perfect tense to show that an action or state started in the past and continues to the present.

Now

Past ◄――► Future

I **have had** my computer
for two months.

Examples	Explanation
Larry Page **has been** interested in computers **for many years.** My sister **has been** a programmer **for three years.**	Use _for_ + an amount of time: _for two months, for three years, for one hour, for a long time,_ etc.
Brin's family **has been** in the U.S. **since 1979.** I **have had** my computer **since March.** Personal computers **have been** popular **since the 1980s.**	Use _since_ with the date, month, year, etc. that the action began.
Brin _has been_ interested in computers _since he_ **was** _a child_. I _have had_ an e-mail account _since I_ **bought** _my computer._	Use _since_ with the beginning of the continuous action or state. The verb in the _since_ clause is simple past.
How long _has_ Brin's family _been_ in the U.S.? **How long** _have_ you _had_ your computer?	Use _how long_ to ask about the amount of time from the past to the present.
Larry Page _has_ **always** _lived_ in the U.S. He _has_ **always** _been_ interested in computers.	We use the present perfect with _always_ to show that an action began in the past and continues to the present.
My grandmother _has_ **never** _used_ a computer. Google _has_ **never** _put_ advertising on its opening page.	We use the present perfect with _never_ to show that something has not occurred from the past to the present.

EXERCISE 6   Fill in the blanks with the missing words.

EXAMPLE   I've known my best friend ___since___ we were in high school.

1. My brother has been in the U.S. _____ 1998.
2. My mother _____ never been in the U.S.
3. How _____ have you been in the U.S.?
4. I've known the teacher since I _____ to study at this college.
5. She's _____ married for two years.
6. She's had the same job _____ ten years.
7. My wife and I _____ known each other since we _____ in elementary school.
8. She' _____ been a student at this college _____ September.
9. I've had my car for three years. _____ long have you _____ your car?
10. I'm interested in art. I' _____ _____ interested in art since I was in high school.
11. _____ always wanted to have my own business.

EXERCISE 7   ABOUT YOU Write true statements using the present perfect with the words given and *for*, *since*, *always*, or *never*. Share your sentences with the class.

EXAMPLES   know My parents have known each other for over 40 years.
             have I've had my car since 2002.
             want I've always wanted to learn English.

1. have _____
2. be _____
3. want _____
4. know _____

EXERCISE 8   ABOUT YOU Make statements with *always*.

EXAMPLE   Name something you've always thought about.
           I've always thought about my future.

1. Name something you've always enjoyed.
2. Name a person you've always liked.

3. Name something you've always wanted to do.

4. Name something you've always wanted to have.

5. Name something you've always been interested in.

EXERCISE 9   ABOUT YOU Make statements with *never*.

EXAMPLE   Name a machine you've never used.
I've never used a fax machine.

1. Name a movie you've never seen.

2. Name a food you've never liked.

3. Name a subject you've never studied.

4. Name a city you've never visited.

5. Name a sport you've never played.

6. Name a food you've never tasted.

EXERCISE 10   ABOUT YOU Write four sentences telling about things you've always done (or been). Share your sentences with the class.

EXAMPLES   I've always cooked the meals in my family.
I've always been lazy.

1. _____

2. _____

3. _____

4. _____

EXERCISE 11   ABOUT YOU Write four sentences telling about things you've never done (or been) but would like to. Share your sentences with the class.

EXAMPLES   I've never studied photography, but I'd like to.
I've never acted in a play, but I'd like to.

1. _____

2. _____

3. _____

4. _____

9.8 | The Simple Present vs. the Present Perfect

Examples	Explanation
a. Larry Page **is** in California.	Sentences (a) refer only to the present.
b. Larry Page **has been** in California since he was in his 20s.	
a. He **loves** computers.	Sentences (b) connect the past to the present.
b. He **has** always **loved** computers.	
a. Google **doesn't have** advertising on its homepage.	
b. Google **has** never **had** advertising on its homepage.	
a. **Do** you **work** at a computer company? Yes, I **do**.	
b. **Have** you always **worked** at a computer company? Yes, I **have**.	

EXERCISE ⓓ Read each statement about your teacher. Then ask the teacher a question beginning with the words given. Include *always* in your question. Your teacher will answer.

EXAMPLE You're a teacher. Have you＿＿＿＿ always been a teacher ＿＿＿＿?
No. I was an accountant before I became a teacher. I've only been a teacher for five years.

1. You teach English. Have you ＿＿＿＿＿＿＿＿＿＿＿＿＿＿＿＿＿＿＿＿
＿＿＿＿＿＿＿＿＿＿＿＿＿＿＿＿＿＿＿＿＿＿＿＿＿＿＿＿＿＿?

2. You work at this college / school. Have you ＿＿＿＿＿＿＿＿＿＿
＿＿＿＿＿＿＿＿＿＿＿＿＿＿＿＿＿＿＿＿＿＿＿＿＿＿＿＿＿＿?

3. You think about grammar. Have you ＿＿＿＿＿＿＿＿＿＿＿＿＿＿
＿＿＿＿＿＿＿＿＿＿＿＿＿＿＿＿＿＿＿＿＿＿＿＿＿＿＿＿＿＿?

4. English is easy for you. Has English ＿＿＿＿＿＿＿＿＿＿＿＿＿＿
＿＿＿＿＿＿＿＿＿＿＿＿＿＿＿＿＿＿＿＿＿＿＿＿＿＿＿＿＿＿?

5. Your last name is ＿＿＿＿＿＿. Has your last name ＿＿＿＿＿＿
＿＿＿＿＿＿＿＿＿＿＿＿＿＿＿＿＿＿＿＿＿＿＿＿＿＿＿＿＿＿?

6. You're interested in languages. Have you ＿＿＿＿＿＿＿＿＿＿＿
＿＿＿＿＿＿＿＿＿＿＿＿＿＿＿＿＿＿＿＿＿＿＿＿＿＿＿＿＿＿?

7. You live in this city. Have you ＿＿＿＿＿＿＿＿＿＿＿＿＿＿＿＿
＿＿＿＿＿＿＿＿＿＿＿＿＿＿＿＿＿＿＿＿＿＿＿＿＿＿＿＿＿＿?

EXERCISE 13 Fill in the blanks with the missing words.

Two students meet by chance in the computer lab.

A: _____Have_____ you _____been_____ in the U.S. for long?
    (examples)

B: No. I _____.
              (1)

A: How _____ _____ you been in the U.S.?
              (2)   (3)

B: I _____ _____ here for about a year.
          (4)   (5)

A: Where do you come from?

B: Burundi.

A: Burundi? I _____
                    (6)

   never _____ of it.
              (7)

   Where is it?

B: It's a small country in Central Africa.

A: Do you have a map? Can you show me where it is?

B: Let's go on the Internet. We can do a search.

A: Did you learn to use a computer in your country?

B: No. When I came here, a volunteer at my church gave me her old
   computer. Before I didn't know anything about computers. I've
   _____ a lot about computers since I came here.
          (8)

A: Oh, now I see Burundi. It's very small. It's near Congo.

B: Yes, it is.

A: Why did you come to the U.S.?

B: My country _____ political problems for many years.
                      (9)

   It wasn't safe to live there. My family left in 1995.

A: So you _____ _____ here since 1995?
                  (10)   (11)

B: No. First we lived in a refugee camp in Zambia.

A: I' _____ never _____ of Zambia either.
              (12)   (13)

   Can we search for it on the Internet?

B: Here it is.

A: You speak English very well. Is English the language of Burundi?

B: No. Kirundi is the official language. Also French. I _____
(14)

_____ French since I was a small child. Where are you
(15)

from?

A: I'm from North Dakota.

B: I _____ never _____ of North Dakota.
(16)   (17)

Is it in the U.S.?

A: Of course. Let's search for an American map on the Internet. Here it is. Winter in North Dakota is very cold. It's cold here too.

B: I don't know how people live in a cold climate. I _____
(18)

never _____ in a cold climate before. I _____
(19)   (20)

always _____ near the Equator.
(21)

A: Don't worry. You'll be OK. You just need warm clothes for the winter.

B: I have class now. I've got to go.

A: I _____ _____ so much about your country
(22)   (23)

in such a short time.

B: It's easy to learn things fast using a computer and a search engine.

9.9 | The Present Perfect vs. the Simple Past

Do not confuse the present perfect with the simple past.	
Examples	**Explanation**
Compare: a. Sergey Brin **came** to the U.S. in 1979. b. Sergey Brin **has been** in the U.S. since 1979. a. Brin and Page **started** Google in 1998. b. Google **has been** popular since 1998.	Sentences (a) show a single action in the past. This action does not continue. Sentences (b) show the continuation of an action or state from the past to the present.
a. When **did** Brin **come** to the U.S.? b. How long **has** Brin **been** in the U.S.?	Question (a) with *when* uses the simple past tense. Question (b) *with how long* uses the present perfect tense.

EXERCISE 14 Fill in the blanks with the simple past or the present perfect of the verb in parentheses ( ).

A: Do you like to surf the Internet?

B: Of course, I do. I ____have had____ my Internet connection since
    (example: have)
    1999, and I love it. A couple of months ago, I _____
                                                    (1 buy)
    a new computer with lots of memory and speed. And last month I
    _____ from a dial-up connection to a cable modem. a
    (2 change)
    dial-up connection to a cable modem. Now I can surf much faster.

A: What kind of things do you search for?

B: Lots of things. I _____ to learn about the stock market,
                    (3 always / want)
    and with the Web, I can start to learn. Last week I _____
                                                        (4 make)
    my first investment in the stock market.

A: Do you ever buy products online?

B: Sometimes I do. Last month, I _____ a great Web site
                                  (5 find)
    where I can download music for 99¢. So far I _____
                                                  (6 download)
    about a hundred songs, and I _____ several CDs.
                                  (7 make)
    My old computer _____ a CD burner, so I'm very
                     (8 not / have)
    happy with my new one.

A: _____ your old computer?
    (9 you/sell)

B: No. It was about eight years old. I just _____ it on top
                                             (10 leave)
    of the garbage dumpster. When I _____ by a few hours
                                     (11 pass)
    later, it was gone. Someone _____ it.
                                 (12 take)

A: Was your new computer expensive?

B: Yes, but I _____ a great deal online.
              (13 get)

A: I _____ my computer for three years, and it seems so
     (14 have)
    old by comparison to today's computers. But it's too expensive to
    buy a new one every year.

B: There's a joke about computers: "When is a computer old?"

A: I don't know. When?

B: As soon as you get it out of the box!

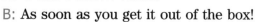

9.10 | The Present Perfect Continuous—An Overview

We use the present perfect continuous to talk about an action that started in the past and continues to the present.

Affirmative	I **have been using** the Internet for two hours.
Negative	You **haven't been working** on your computer all day.
Question	**Have** you **been surfing** the web for great deals?

GENEALOGY

Before You **Read**

1. Do you think it's important to know your family's history? Why or why not?

2. What would you like to know about your ancestors?

 Read the following article. Pay special attention to the present perfect and the present perfect continuous tenses.

Did You **Know?**

Family history is the second most popular hobby in the U.S. after gardening.

In the last 30 years, genealogy **has become** one of America's most popular hobbies. If you type *genealogy* in a search engine, you can find about 16 million hits. If you type *family history*, you will get about 10 million hits. The percentage of the U.S. population interested in family history **has been increasing** steadily. Forty-five percent of Americans in 1996 stated they were interested in genealogy. In 2000, that number rose to 60 percent according to a national survey. This increase probably has to do with the ease of searching on the Internet.

The number of genealogy Web sites **has been growing** accordingly as people ask themselves: Where does my family come from? How long **has** my family **been** in the U.S.? Why did they come here? How did they come here? What kind of people were my ancestors?

Genealogy is a lifelong hobby for many. The average family historian **has been doing** genealogy for 14 years, according to a recent study. Most family historians are over 40. Cyndi Howells, from Washington State, quit her job in 1992 and **has been working** on her family history ever since. She **has created** a Web site to help others with their search. Her Web site has over 99,000 resources. Since its start in 1996, her Web site **has had** over 22 million visitors and more than 32 million page hits each month. Cyndi **has** also **been giving** lectures all over the country to genealogy groups. Cyndi's Web site **has won** an award three times for the best genealogy site on the Web.

While the Internet **has made** research easier for amateur genealogists, it is only the beginning for serious family historians. Researchers still need to go to courthouses and libraries to find public records, such as land deeds[5], obituaries[6], wedding notices, and tax records. Another good source of information is the U.S. Census. Early census records are not complete, but since the mid-1800s, the U.S. Census **has been keeping** detailed records of family members, their ages, occupations, and places of birth.

Are you interested in knowing more about your ancestors and their stories, their country or countries, and how you fit into the history of your family? Maybe genealogy is a good hobby for you.

9.11 | The Present Perfect Continuous—Forms

Subject	have/has	been	Present Participle	Complement
I	have	been	using	the Internet for two hours.
We	have	been	reading	about search engines.
You	have	been	studying	computers.
They	have	been	living	in California.
He	has	been	writing	since one o'clock.
She	has	been	surfing	the Internet all day.
It	has	been	raining	all day.

Language Notes:
1. To form the negative, put *not* between *have* or *has* and *been*.
   You **have *not* been** listening.
   She **has*n't* been** working hard.
2. To form the question, reverse the subject and *have/has.*
   Has she been using her new computer?
   How long **have they** been living in the U.S.?

---

[5] A *land deed* is a document that shows who the owner of the land is.

[6] *Obituaries* are death notices posted in the newspaper.

9.12 | The Present Perfect Continuous—Statements and Questions

Compare affirmative statements and questions

Wh–Word	have/has	Subject	have/has	Been + Verb -ing	Complement	Short Answer
		Cyndi	**has**	**been working**	on her family history.	
	Has	she		**been working**	on her Web site?	Yes, she has.
How long	has	she		**been working**	on her Web site?	Since 1992.

Compare negative statements and questions

Wh–Word	haven't/hasn't	Subject	haven't/hasn't	Been + Verb-ing	Complement
		They	**haven't**	**been using**	the public library.
Why	**haven't**	they		**been using**	the public library?

EXERCISE **15**   Fill in the blanks with the present perfect continuous form of the verb in parentheses ( ).

EXAMPLE   How long ___has___ Cyndi ___been managing___ a genealogy Web site?

(*example: manage*)

1. Interest in genealogy _____.

(*grow*)

2. Cyndi _____ on her family history since 1992.

(*work*)

3. Cyndi _____ all over the U.S. to

(*lecture*)

genealogy groups.

4. The number of genealogy Web sites _____.

(*increase*)

5. How long _____ the U.S. Census

(*keep*)

_____ records?

6. _____ you _____ a family tree

(*make*)

for your family?

7. People_____ the Internet to do family

(*use*)

research for about ten years.

8. My family_____ in the U.S. for

(*not/live*)

many generations.

9.13 | The Present Perfect Continuous—Use

We use the present perfect continuous tense to show that an action or state started in the past and continues to the present.

Now

Past ◄─────────────────────────── ┊ ─────────────────────────► Future

> He **has been living** in the U.S. since 1979.

Examples	Explanation
Cyndi **has been working** on her family tree since 1992. Sergey Brin **has been living** in the U.S. for more than 25 years.	We use *for* and *since* to show the time spent at an activity.
He **has been living** in the U.S. since 1979. OR He **has lived** in the U.S. since 1979.	With some verbs (*live, work, study, teach,* and *wear*) we can use either the present perfect or the present perfect continuous with actions that began in the past and continue to the present. The meaning is the same.
My father *is working* on the family tree right now. He **has been working** on it since 9 o'clock this morning.	If the action is still happening, use the present perfect continuous, not the present perfect.
Google **has become** one of the most popular search engines. I **have had** my computer for three months.	We do not use the continuous form with nonaction verbs. See below for a list of nonaction verbs.
I **have** always **taken** computer courses. My grandmother **has** never **used** a computer.	Do not use the continuous form with *always* and *never*.

Action: I **having been thinking** *about* doing a family tree.
Nonaction: I **have** always **thought** *that* genealogy is an interesting hobby.
	Think can be an action or nonaction verb, depending on its meaning.
	    Think about = action verb
	    Think that = nonaction verb

Nonaction: Some people **have had** a lot of success in locating information. Action: We **have been having** a hard time locating information about our ancestors.	*Have* is usually a nonaction verb. However, *have* is an action verb in these expressions: *have experience, have a hard time, have a good time, have difficulty,* and *have trouble.*

Nonaction verbs:

like	know	see	cost
love	believe	smell	own
hate	think (that)	hear	have (for possession)
want	care (about)	taste	become
need	understand	feel	
prefer	remember	seem	

EXERCISE 16 ABOUT YOU Write **true** statements using the present perfect continuous with the words given and *for* or *since*. Share your sentences with the class.

EXAMPLE **work** _My brother has been, working as a waiter for six years._

1. study English _____
2. work _____
3. live _____
4. use _____
5. study _____

EXERCISE 17 ABOUT YOU Read aloud each of the following present tense questions. Another student will answer. If the answer is *yes*, add a present perfect continuous question with "*How long have you ...?*"

EXAMPLE Do you play a musical instrument?

A: Do you play a musical instrument?
B: Yes. I play the piano.
A: How long have you been playing the piano?
B: I've been playing the piano since I was a child.

1. Do you drive?
2. Do you work?
3. Do you use the Internet?
4. Do you wear glasses?
5. Do you play a musical instrument?

EXERCISE 18 Ask the teacher questions with "*How long...?*" and the present perfect continuous form of the words given. The teacher will answer your questions.

EXAMPLE speak English

A: How long have you been speaking English?
B: I've been speaking English all[7] my life.

1. teach English
2. work at this school
3. live in this city

4. use this book
5. live at your present address

[7] We do not use the preposition *for* before *all*.

EXERCISE 19   Fill in the blanks in the following conversations. Answers may vary.

EXAMPLE   A: Do you wear glasses?

B: Yes, I _____do_____ .

A: How long _____have_____ you _____been wearing_____ glasses?

B: I _'ve been wearing_ glasses since I _____was_____ in high school.

1. A: Are you working on your family history?

   B: Yes, I am.

   A: How long _____ you _____ on
      your family history?

   B: I _____ on it for about ten years.

2. A: Is your sister surfing the Internet?

   B: Yes, she _____ .

   A: How long _____ she _____
      surfing the Internet?

   B: Since she woke up this morning!

3. A: Does your father live in the U.S.?

   B: Yes, he _____ .

   A: How long _____ he been _____
      in the U.S.?

   B: He _____ in the U.S. since he
      _____ 25 years old.

4. A: Are you studying for the test now?

   B: Yes, I _____ .

   A: How long _____ for the test?

   B: For _____ .

5. A: Is your teacher teaching you the present perfect lesson?

   B: Yes, he _____ .

   A: _____ long _____
      you this lesson?

   B: Since _____ .

6. A: Are they using the computers now?

    B: Yes, _____.

    A: How long _____ them?

    B: _____ they started to write

    their compositions.

7. A: _____ you using the Internet?

    B: Yes, I _____.

    A: How _____?

    B: _____ for two hours.

8. A: _____ your grandparents live in the U.S.?

    B: Yes, they _____.

    A: How _____ in the U.S.?

    B: Since they _____ born.

9. A: Is she studying her family history?

    B: Yes, she _____.

    A: How long _____?

    B: Since she _____.

9.14 | The Present Perfect with Repetition from Past to Present

We use the present perfect to talk about the repetition of an action in a time period that includes the present. There is a probability that this action will occur again.

```
                                    Now
Past ◄───────────────────────────────┆─ ─ ─ ─ ─ ─ ─ ─ ─ ─ ─ ─ ─► Future
            ┌─────────────────────────┐
            │ Cyndi has won three     │
            │ awards so far.          │
            └─────────────────────────┘
```

Examples	Explanation
a. Cyndi **has won** three awards. b. Cyndi's Web site **has had** over 22 million visitors.	a. Cyndi may win another award. b. Cyndi's Web site will probably get more visitors.
So far, Brin and Page **haven't changed** their lifestyles. We **have completed** eight lessons in this book up to now.	Adding the words *so far* and *up to now* indicate that we are counting up to the present.
How many "hits" **has** Cyndi's Web site **had** this month? How much information **have** you **gotten** from her Web site so far?	We can ask a question about repetition with *how many* and *how much*.
How many times **have** you **checked** your e-mail today? I **haven't checked** my e-mail **at all** today.	To indicate zero times, we use a negative verb + *at all*. But there is a probability that this action may happen.
Compare: a. Google **had** 10,000 searches a day in 1998. b. Google **has had** billions of searches since1998. a. Cyndi's list **appeared** for the first time in 1996. b. Many new genealogy Web sites **have appeared** in the last ten years.	a. We use the simple past with a time period that is finished or closed: *1998, 50 years ago,* etc. b. We use the present perfect in a time period that is open. There is a probability of more repetition.

Language Note:
Do not use the continuous form for repetition.
Right: I **have checked** my e-mail three times today.
Wrong: I *have been checking* my e-mail three times today.

EXERCISE **20** ABOUT YOU Ask *a yes / no* question with *so far* or *up to now* and the words given. Another student will answer.

EXAMPLE you / come to every class.
A: Have you come to every class so far?
B: Yes, I have.
   OR
B: No, I haven't. I've missed three classes.

1. we / have any tests

2. this lesson / be difficult

3. the teacher / give a lot of homework

4. you / understand all the explanations

5. you / have any questions about this lesson

EXERCISE **21** ABOUT YOU Ask a question with *"How many...?"* and the words given. Talk about this month. Another student will answer.

EXAMPLE times / go to the post office
A: How many times have you gone to the post office this month?
B: I've gone to the post office once this month.
         OR
I haven't gone to the post office at all this month.

1. letters / write

2. times / eat in a restaurant

3. times / get paid

4. long-distance calls / make

5. books / buy

6. times / go to the movies

7. movies / rent

8. times / cook

EXERCISE **22** ABOUT YOU Write four questions to ask another student or your teacher about repetition from the past to the present. Use *how much* or *how many*. The other person will answer.

EXAMPLES How many cities have you lived in?
How many English courses have you taken at this college?

1. _____

2. _____

3. _____

4. _____

9.15 | The Simple Past vs. the Present Perfect with Repetition

We use the present perfect with repetition in a present time period. There is probability of more repetition. We use the simple past with repetition in a past time period. There is no possibility of any more repetition during that period.

Examples	Explanation
How many hits **has** your Web site **had** today? It **has had** over 100 hits today. How many times **have** you **been** absent this semester? I've **been** absent twice so far.	To show that there is possibility for more repetition, use the present perfect. In the examples on the left, *today* and *this semester* are not finished. *So far* indicates that the number given may not be final.
Last month my Web site **had** 5,000 hits. How many times **were** you absent last semester?	To show that the number is final, use the simple past tense and a past time expression. *Yesterday, last week, last year, last semester,* etc. are finished. The number is final.
a. Brin and Page **have added** new features to Google over the years. b. Before she died, my grandmother **added** many details to our family tree.	a. Brin and Page are still alive. They can (and probably will) add new features to Google in the years to come. b. Grandmother is dead. The number of details she added is final.
Compare: a. I **have checked** my e-mail twice today. b. I **checked** my e-mail twice today. a. She **has gone** to the library to work on her family tree five times this month. b. She **went** to the library to work on her family tree five times this month.	With a present time expression (such as *today*, *this week*, *this month*, etc.), you may use either the present perfect or the simple past. In sentences (a), the number may not be final. In sentences (b), the number seems final.
Compare: a. In the U.S., I **have had** two jobs. b. In my native country, I **had** five jobs. a. In the U.S., I **have lived** in three apartments so far. b. In my native country, I **lived** in two apartments.	a. To talk about your experiences in this phase of your life, you can use the present perfect tense. b. To talk about a closed phase of your life, use the simple past tense. For example, if you do not plan to live in your native country again, use the simple past tense to talk about your experiences there.

EXERCISE 23 ABOUT YOU Fill in the blanks with the simple past or the present perfect to ask a question. A student from another country will answer.

EXAMPLES How many cars ___have you owned___ in the U.S.?
I've owned two cars in the U.S.

How many cars ___did you own___ in your country?
I owned only one car in my country.

1. How many apartments _____ in your country?
2. How many apartments _____ in the U.S.?
3. How many schools _____ in your country?
4. How many schools _____ in the U.S.?
5. How many jobs _____ in the U.S.?
6. How many jobs _____ in your country?

9.16 | The Present Perfect with Indefinite Past Time

We use the present perfect to refer to an action that occurred at an indefinite time in the past that still has importance to the present situation. Words that show indefinite time are: *ever, yet,* and *already*.

Now

Past ◄─────────── | ─────────── ► Future

Have you ever **used** Google?

Examples	Explanation

Have you *ever* **noticed** that Google doesn't have ads on its opening page?
  Yes, I **have**.
Have you *ever* **studied** your family history?
  No, I never **have**.
Have you *ever* "**googled**" your name?
  No, I haven't.
	A question with *ever* asks about any time between the past and the present. Put *ever* between the subject and the main verb.

Has Larry Page **gotten** his degree from Stanford University *yet*?
  No, not *yet*.
Have Larry and Sergey **become** billionaires *yet*?
  Yes, they have.
Have you **read** the story about genealogy *yet*?
  Yes, I *already* have.
	Yet and *already* refer to an indefinite time in the near past. There is an expectation that an activity took place a short time ago.

The success of Google over its rivals **has shown** that users don't want to see a lot of ads. Google **has added** many new features to its Web site. Cyndi Howells **has created** a very useful Web site for family historians.	We can use the present perfect to talk about the past without any reference to time. The time is not important or not known or imprecise. Using the present perfect, rather than the simple past, shows that the past is relevant to a present situation.

EXERCISE **24**   ABOUT YOU Answer the following questions with: *Yes, I have; No, I haven't;* or *No, I never have.*

1. Have you ever "googled" your own name?
2. Have you ever researched your family history?
3. Have you ever made a family tree?
4. Have you ever used the Web to look for a person you haven't seen in a long time?
5. Have you ever added hardware to your computer?
6. Have you ever downloaded music from the Internet?
7. Have you ever used a search engine in your native language?
8. Have you ever sent photos by e-mail?
9. Have you ever received a photo by e-mail?
10. Have you ever bought anything from a Web site?
11. Have you ever built a computer?
12. Has your computer ever had a virus?

EXERCISE **25**   ABOUT YOU Answer the questions with: *Yes, I have; Yes, I already have;* or *Not yet.*

1. Have you eaten lunch yet?
2. Have you finished lesson eight yet?
3. Have you done today's homework yet?
4. Have you paid this month's rent yet?
5. Have you learned the names of all the other students yet?
6. Have you visited the teacher's office yet?
7. Have you done Exercise 22 yet?
8. Have you learned the present perfect yet?
9. Have you learned all the past participles yet?

9.17 | Answering a Present Perfect Question

We can answer a present perfect question with the simple past tense when a specific time is introduced in the answer. If a specific time is not known or necessary, we answer with the present perfect.

Examples	Explanation
Have you ever **used** Google? Answer A: Yes. I'**ve used** Google many times. Answer B: Yes. I **used** Google a few hours ago.	Answer A, with *many times*, shows repetition at an indefinite time. Answer B, with *a few hours ago*, shows a specific time in the past.
Have you ever **heard** of Larry Page? Answer A: No. I'**ve never** heard of him. Answer B: Yes. We **read** about him yesterday.	Answer A, with *never*, shows continuation from past to present. Answer B, with *yesterday*, shows a specific time in the past.
Have you **done** your homework yet? Answer A: Yes. I'**ve done** it already. Answer B: Yes. I **did** it this morning.	Answer A, with *already*, is indefinite. Answer B, with *this morning*, shows a specific time.
Have Brin and Page **become** rich? Answer A: Yes, they have. Answer B: Yes. They **became** rich before they were 30 years old.	Answer A is indefinite. Answer B, with *before they were 30 years old*, is definite.

EXERCISE 26   ABOUT YOU Ask a question with *"Have you ever...?"* and the present perfect tense of the verb in parentheses ( ). Another student will answer. To answer with a specific time, use the past tense. To answer with a frequency response, use the present perfect tense. You may work with a partner.

EXAMPLES   (go) to the zoo

A: Have you ever gone to the zoo?
B: Yes. I've gone there many times.

(go) to Disneyland

A: Have you ever gone to Disneyland?
B: Yes. I went there last summer.

1. (work) in a factory
2. (lose) a glove
3. (run) out of gas[8]
4. (fall) out of bed

[8] *To run out of gas* means to use all the gas in your car while driving.

5. (make) a mistake in English grammar

6. (tell) a lie

7. (eat) raw[9] fish

8. (study) calculus

9. (meet) a famous person

10. (go) to an art museum

11. (stay) up all night

12. (break) a window

13. (get) locked out[10] of your house or car

14. (see) a French movie

15. (go) to Las Vegas

16. (travel) by ship

17. (be) in love

18. (write) a poem

EXERCISE **27** ABOUT YOU Write five questions with *ever* to ask your teacher. Your teacher will answer.

EXAMPLES Have you ever gotten a ticket for speeding?

Have you ever visited Poland?

1. _____

2. _____

3. _____

4. _____

5. _____

EXERCISE **28** ABOUT YOU Ask a student from another country questions using the words given. The other student will answer.

EXAMPLE your country / have a woman president

A: Has your country ever had a woman president?

B: Yes, it has. We had a woman president from 1975 to 1979.

1. your country / have a civil war

2. your country's leader / visit the U.S.

3. an American president / visit your country

4. your country / have a woman president

5. you / go back to visit your country

6. there / be an earthquake in your hometown

---

[9] *Raw* means not cooked.

[10] *To get locked out* of your house means that you can't get in because you do not have keys with you to get inside.

EXERCISE **29** ABOUT YOU Ask a student who has recently arrived in this country if he or she has done these things yet.

EXAMPLE **buy a car**
A: Have you bought a car yet?
B: Yes, I have. OR No, I haven't. OR I bought a car last month.

1. find a job
2. make any American friends
3. open a bank account
4. save any money
5. buy a car

6. write to your family
7. get a credit card
8. buy a computer
9. get a telephone

EXERCISE **30** *Combination Exercise.* Fill in the blanks with the correct tense of the verb in parentheses ( ). Also fill in other missing words.

A: Your Spanish is a little different from my Spanish. Where are you from?

B: I'm from Guatemala.

A: How _____long_____ ____have you been____ here?
          (example)   *(example: you/be)*

B: I _____ here for about six months. Where are
        (1 only/be)

you from?

A: Miami. My family comes from Cuba. They _____
                                          (2 leave)

Cuba in 1962. after the revolution. I _____ born in
                                         (3 be)

the U.S. I'm starting to become interested in my family's history.

I _____ several magazine articles about genealogy
  (4 read)

so far. It's fascinating. Are you interested in your family's history?

B: Of course I am. I _____ interested in it _____
             (5 be)   *(6)*

a long time. I _____ on a family tree for many years.
         (7 work)

A: When _____?
              (8 yon/start)

B: I _____ when I _____ 16 years old.
    (9 start)   *(10 be)*

Over the years, I _____ a lot of interesting
                       (11 find)

information about my family. Some of my ancestors were Mayans and some were from Spain and France. In fact, my great-great grandfather was a Spanish prince.

A: How _____ all that information?
               (12 you/find)

B: I _____ the Internet a lot. I _____
      (13 used)   *(14 also/go)*

to many libraries to get more information.

A: _____ to Spain or France to look at records
     (15 ever/go)

there?

B: Last summer I _____ to Spain, and I _____
            (16 go)   *(17 find)*

a lot of information while I was there.

A: How many ancestors _____ so far?
                      (18 you/find)

B: So _____ I _____ about 50,
          (19)   *(20 find)*

but I'm still looking.

A: How can I get started?

B: There's a great Web site called "Cyndi's list." I'll give you the Web address, and you can get started there.

1. Compare the present perfect and the simple past.

Present Perfect	Simple Past
A. The action of the sentence began in the past and includes the present:	A. The action of the sentence is completely past:

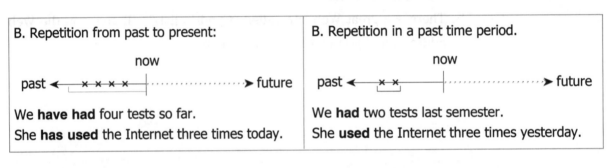

Present Perfect	Simple Past
My father **has been** in the U.S. since 1992.	My father **came** to the U.S. in 1992.
My father **has had** his job in the U.S. for many years.	My father **was** in Canada for two years before he came to the U.S.
How long **have** you **been** interested in genealogy?	When **did** you **start** your family tree?
I'**ve** always **wanted** to learn more about my family's history.	When I was a child, I always **wanted** to spend time with my grandparents.

B. Repetition from past to present:	B. Repetition in a past time period.

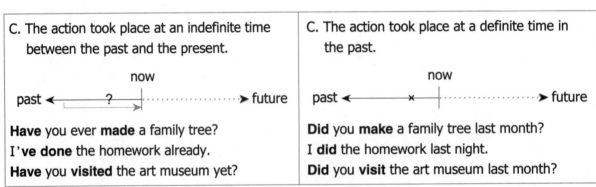

Present Perfect	Simple Past
We **have had** four tests so far.	We **had** two tests last semester.
She **has used** the Internet three times today.	She **used** the Internet three times yesterday.

C. The action took place at an indefinite time between the past and the present.	C. The action took place at a definite time in the past.

Present Perfect	Simple Past
Have you ever **made** a family tree?	**Did** you **make** a family tree last month?
I'**ve done** the homework already.	I **did** the homework last night.
Have you **visited** the art museum yet?	**Did** you **visit** the art museum last month?

2. Compare the present perfect and the present perfect continuous.

Present Perfect	Present Perfect Continuous

A. A continuous action (nonaction verbs)
   I **have had** my car for five years.
B. A repeated action
   Cyndi's Web site **has won** several awards.
C. Question with *how many*
   How many times **have** you **gone** to New York?
D. An action that is at an indefinite time, completely in the past.
   Cyndi **has created** a Web site.
	A. A continuous action (action verbs)
	   I **'ve been driving** a car for 20 years.
	B. A nonstop action
	   The census **has been keeping** records since the 1880s.
	C. Question with *how long*
	   How long **has** he **been living** in New York?
	D. An action that started in the past and is still happening.
	   Cyndi **has been working** on her family history since 1992.

EDITING ADVICE

1. Don't confuse the *–ing* form and the past participle.

       taking
She has been ~~taken~~ a test for two hours.

       given
She has ~~giving~~ him a present.

2. Use the present perfect, not the simple present, to describe an action or state that started in the past and continues to the present.

      had
He has∧ a car for two years.

    have   ed
How long ~~do~~ you work∧ in a factory?

3. Use *for*, not *since*, with the amount of time.

       for
I've been studying English ~~since~~ three months.

4. Use the simple past, not the present perfect, with a specific past time.

      came
He ~~has come~~ to the U.S. five months ago.

      did
When ~~have~~ you come to the U.S.?

5. Use the simple past, not the present perfect, in a *since* clause.

           came
He has learned a lot of English since he ~~has come~~ to the U.S.

6. Use correct word order. Put the adverb between the auxiliary and the main verb.

> never seen
> He has ~~seen never~~ a French movie.

> ever gone
> Have you ~~gone ever~~ to France?

7. Use correct word order in questions.

> have you
> How long ~~you have~~ been a teacher?

8. Use *yet* for negative statements; use *already* for affirmative statements.

> yet
> I haven't eaten dinner ~~already~~.

9. Don't forget the verb *have* in the present perfect (continuous).

> have
> I ∧ been living in New York for two years.

10. Don't forget the *-ed* of the past participle.

> ed
> He's listen ∧ to that CD many times.

11. Use the present perfect, not the continuous form, with *always*, *never*, *yet*, *already*, *ever*, and *how many*.

> gone
> How many times have you ~~been going~~ to Paris?

> visited
> I've never ~~been visiting~~ Paris.

12. Don't use *time* after *how long*.

> How long ~~time~~ have you had your job?

LESSON 9 TEST/REVIEW

PART **1**   Find the mistakes with the underlined words, and correct them. Not every sentence has a mistake. If the sentence is correct, write *C*.

> had
> EXAMPLES   I have ∧ my car for six years.
>
> We've always wanted to learn English. C

1. Since I've come to the U.S., I've been studying English.
2. Have you ever eating Chinese food?
3. How long you've been in the U.S.?

4. Have you <u>gone ever</u> to Canada?

5. I've <u>know</u> my best friend since I was a child.

6. She's a teacher. She's been a teacher <u>since</u> ten years.

7. <u>I never gone</u> to Mexico.

8. <u>How long time</u> has your father been working as an engineer?

9. Has he ever gone to Paris? Yes, he <u>went</u> to Paris last year.

10. He works in a restaurant. <u>He been working</u> there since 1995.

11. Have you ever <u>study</u> biology?

12. Have they finished the test <u>yet</u>?

13. She's done the homework <u>yet</u>.

PART **2** Fill in the blanks with the simple past, the present perfect, or the present perfect continuous form of the verb in parentheses ( ). In some cases, more than one answer is possible.

Conversation 1

A: _____Have_____ you ever _____studied_____ computer programming?
   (example: study)

B: Yes. I _____ it in college. And I _____
           (1 study)   *(2 work)*

as a programmer for five years. But my job is boring.

A: _____ you ever _____ about changing jobs?
                                   (3 think)

B: Yes. Since I _____ a child, I _____ to be an
                 (4 be)   *(5 always/want)*

actor. When I was in college, I _____ in a few plays, but
                                        (6 be)

since I _____, I _____ time
         (7 graduate)   *(8 not/have)*

to act.

Conversation 2

A: How long _____ in the U.S.?
              (1 you/be)

B: For about two years.

A: _____ your life _____ a lot since
                                   (2 change)

you _____ to the U.S?
      (3 come)

B: Oh, yes. Before I _____ here, I _____

(4 come)   (5 live)

with my family. Since I came here, I _____ alone.

(6 live)

A: _____ in the same apartment in this city?

(7 always/live)

B: No. I _____ three times so far. And I plan to

(8 move)

move again at the end of the year.

A: Do you plan to have a roommate?

B: Yes, but I _____ one yet.

(9 not/find)

PART **3** Fill in the blanks with the simple present, the simple past, the present perfect, or the present perfect continuous form of the verb in parentheses ( ). In some cases, more than one answer is possible.

Paragraph 1

I _____ the Internet every day. I

(1 use)

_____ it for three years. I _____

(2 use)   (3 start)

to use it when I _____ interested in genealogy. I

(4 become)

_____ on my family tree for three years. Last

(5 work)

month, I _____ information about my father's

(6 find)

ancestors. My grandfather _____ with us now and likes

(7 live)

to tell us about his past. He _____ born in Italy, but he

(8 be)

_____ here when he was very young, so he

(9 come)

_____ here most of his life. He doesn't remember much

(10 live)

about Italy. I _____ any information about my

(11 not / find)

mother's ancestors yet.

Paragraph 2

I _____ to the U.S. when a war
                    (1 come)

_____ out in my country. I _____
        (2 break)   *(3 live)*

in the U.S. for five years. At first, everything _____
                                                        (4 be)

very hard for me. I _____ any English when I
                        (5 not/know)

_____ . But I _____ English for the
      (6 arrive)   *(7 study)*

past five years, and now I _____ it pretty well.
                                (8 speak)

I _____ my college education yet, but I plan to
        (9 not/start)

next semester.

EXPANSION ACTIVITIES

Classroom Activities

1. Form a group of between 4 and 6 students. Find out who in your group
   has done each of these things. Write that person's name in the blank.

   a. _____ has made a family tree.
   b. _____ has found a good job.
   c. _____ has been on a ship.
   d. _____ has never eaten Mexican food.
   e. _____ hasn't done today's homework yet.
   f. _____ has never seen a French movie.
   g. _____ has taken a trip to Canada.
   h. _____ has acted in a play.
   i. _____ has gone swimming in the Pacific Ocean.
   j. _____ has flown in a helicopter.
   k. _____ has served in the military.
   l. _____ has worked in a hotel.
   m. _____ has never studied chemistry.
   n. _____ has taken the TOEFL test.
   o. _____ has just gotten a "green card."

2. Draw your family tree for the past three generations, if you can. Why do you think so many people are interested in genealogy? What is valuable about finding your family's history?

Write About it

1. Write a composition about one of the following:
   How your life has changed *(choose one)*:
   a. since you came to the U.S.
   b. since you got married
   c. since you had a baby
   d. since you started college
   e. since you graduated from high school

2. Write about an interesting member of your family. What has he or she done that you think is interesting?

Outside Activity

Interview an American who has relatives who have been in the U.S. for several generations. Does this person know the stories of his or her ancestors and their native countries? What is something interesting you discovered from this interview?

Internet Activities

1. On the Internet, find Cyndi Howell's genealogy Web site. Find out about people who have the same last name as yours.

2. Type the word *genealogy* at a search engine. How many Web sites did you find?

3. Go to a search engine and type in *Larry Page, Sergey Brin*. Find an interesting fact about one of them that you didn't know. Bring it to class.

Additional Activities at http://elt.thomson.com/gic

GRAMMAR

Gerunds
Infinitives

CONTEXT: Finding a Job

Finding a Job
Tips on Writing a Résumé
Rita's Story

10.1 | Gerunds—An Overview

To form a gerund, we use the *-ing* form of a verb (*finding, learning, eating, running*). A *gerund phrase* is a gerund+a noun phrase (*finding a job, learning English*). A gerund (phrase) can appear in several positions in a sentence.

Examples	Explanation
Finding a job is hard.	• The gerund is the subject.
I don't enjoy **talking** about myself.	• The gerund is the object.
I thought about **changing** my career.	• The gerund is the object of the preposition.
I got information **by talking** with my counselor.	• The gerund is part of an adverbial phrase.
I like to **go shopping**.	• The gerund is in many expressions with *go*.
Not having a job is frustrating. You can impress the boss by **not being** late.	We can put *not* in front of a gerund to make it negative.

FINDING A JOB

Before You Read

1. Have you ever had a job interview in this city?

2. What is your profession or job? What profession or job do you plan to have in the future?

 Read the following article. Pay special attention to gerunds.

Finding a job in the United States takes specific skills. The following advice will help you find a job.

• Write a good résumé. Describe your accomplishments.[1] Avoid **including** unnecessary information. Your résumé should be one page, if possible.

[1]*Accomplishments* are the unusual good things you have done, such as awards you have won or projects you have successfully managed.

- Find out about available jobs. One way is by **looking** in the newspaper or on the Internet. Another way is by **networking**. **Networking** means **exchanging** information with anyone you know—family, friends, neighbors, classmates, former coworkers, professional groups—who might know of a job. These people might also be able to give you insider information about a company, such as who is in charge and what it is like to work at their company. According to an article in the *Wall Street Journal*, 94 percent of people who succeed in **finding** a job say that **networking** was a big help.

- Practice the interview. The more prepared you are, the more relaxed you will feel. If you are worried about **saying** or **doing** the wrong thing, practice will help.

- Learn something about the company. You can find information by **going** to the company's Web site. **Getting** information takes time, but it pays off.

You can get help in these skills—**writing** a résumé, **networking, preparing** for an interview, **researching** a company—by **seeing** a career counselor. Most high schools and colleges have one who can help you get started.

    Finding a job is one of the most difficult jobs. Some people send out hundreds of résumés and go on dozens of interviews before **finding** a job. And it isn't something you do just once or twice in your lifetime. For most Americans, **changing** jobs many times in a lifetime is not uncommon.

Tips for Getting a Job

Preparation:
1. Learn about the organization and have a specific job or jobs in mind.
2. Review your résumé.
3. Practice an interview with a friend or relative.
4. Arrive at least 15 minutes before the scheduled time of your interview.

Personal appearance:
1. Be well groomed[2] and dress appropriately.
2. Do not chew gum.

The interview:
1. Relax and answer each question concisely.
2. Use good manners. Shake hands and smile when you meet someone.
3. Be enthusiastic. Tell the interviewer why you are a good candidate for the job.
4. Ask questions about the position and the organization.
5. Thank the interviewer when you leave and in writing as a follow-up.

Information to bring to an interview:
1. Social Security card.
2. Government-issued identification (driver's license).
3. Résumé or application. Include information about your education, training, and previous employment.
4. References. Employers typically require three references. Get permission before using anyone as a reference. Make sure that each will give you a good reference. Avoid using relatives as references.

[2]When you are *well groomed*, your appearance is neat and clean.

10.2 | Gerund as Subject

Examples	Explanation
Gerund Phrase **Finding a good job** takes time. **Writing a résumé** isn't easy.	We can use a gerund or gerund phrase as the subject of the sentence.
Exchanging ideas with friends **is** helpful. **Visiting** company Web sites **takes** time.	A gerund subject takes a singular verb.
Not preparing for an interview could have a bad result.	We can put *not* in front of a gerund to make it negative.

EXERCISE **1**   The following things are important before a job interview. Make a sentence with each one, using a gerund phrase as the subject.

EXAMPLE   **get a good night's sleep**

Getting a good night's sleep will help you feel rested and alert for an interview.

1. **take a bath or shower**

   _____

2. **select serious-looking clothes**

   _____

3. **prepare a résumé**

   _____

4. **check your résumé carefully**

   _____

5. **get information about the company**

   _____

6. **prepare answers to possible questions**

   _____

EXERCISE **2**   Complete each statement with a gerund (phrase) as the subject.

EXAMPLE   _____ Learning a foreign language _____ takes a long time.

1. _____ is one of the most difficult jobs.
2. _____ is one of the best ways to find a job.
3. _____ is not permitted in this classroom.
4. _____ is difficult for a foreign student.
5. _____ takes a long time.
6. _____ is not polite.

7. _____ makes me feel good.

8. _____ makes me nervous.

9. _____ scares me.

10. _____ is against the law.

EXERCISE **3**   ABOUT YOU In preparing for an interview, it is good to think about the following questions. Answer these questions. Use a gerund in some of your answers, but do NOT try to use a gerund in every answer. It won't work. Give a lot of thought to your answers and compare them with your classmates' answers.

EXAMPLES   What are your strengths?

Working with others; learning quickly; thinking fast in difficult situations

What are your strong and weak subjects in school?

I'm strong in math. I'm weak in history.

1. What are your strengths?

   _____

2. What are some of your weaknesses?

   _____

3. List your accomplishments and achievements. (They can be achievements in jobs, sports, school, etc.)

   _____

4. What are your interests?

   _____

5. What are your short-term goals?

   _____

6. What are your long-term goals?

   _____

7. What are things you like? Think about personalities, tasks, environments, types of work, and structure.

   _____

8. What are some things you dislike? Think about personalities, tasks, environments, types of work, and structure.

   _____

9. Why should we hire you?

   _____

Write a list of personal behaviors during an interview that would hurt your chances of getting a job. You may work with a partner or in a small group.

EXAMPLES   *Chewing gum during the interview looks bad.*

Not looking directly at the interviewer can hurt your chances.

1. _____

2. _____

3. _____

4. _____

5. _____

10.3 | Gerund After Verb

Some verbs are commonly followed by a gerund (phrase). The gerund (phrase) is the object of the verb.

Examples	Explanation
Have you considered **going** to a job counselor? Do you appreciate **getting** advice? You can discuss **improving** your skills. You should practice **answering** interview questions.	The verbs below can be followed by a gerund: admit discuss mind put off appreciate dislike miss quit avoid enjoy permit recommend can't help finish postpone risk consider keep practice suggest
I have many hobbies. I like to **go fishing** in the summer. I **go skiing** in the winter. I like indoor sports too. I **go bowling** once a month.	*Go*+gerund is used in many idiomatic expressions. go boating go jogging go bowling go sailing go camping go shopping go dancing go sightseeing go fishing go skating go hiking go skiing go hunting go swimming
a. I don't **mind** *wearing* a suit to work. b. Don't **put off** *writing* your résumé. Do it now. c. I have an interview tomorrow morning. I **can't help** *feeling* nervous.	a. *I mind* means that something bothers me. *I don't mind* means that something is OK with me; it doesn't bother me. b. *Put off* means postpone. c. *Can't help* means to have no control over something.

EXERCISE **5**   ABOUT YOU Fill in the blanks with an appropriate gerund (or noun) to complete these statements. Share your answers with the class.

EXAMPLE   I don't mind ___shopping for food___, but I do[3] mind ___cooking it___ .

1. I usually enjoy _____ during the summer.
2. I don't enjoy _____ .
3. I don't mind _____, but I do mind

_____ .

4. I appreciate _____ from my friends.
5. I need to practice _____ if I want to improve.
6. I often put off _____ .
7. I need to keep _____ if I want to be successful.
8. I should avoid _____ if I want to improve my health.
9. I miss _____ from my hometown.

EXERCISE **6**   ABOUT YOU Make a list of suggestions and recommendations for a tourist who is about to visit your hometown. Read your list to a partner, a small group, or the entire class.

EXAMPLES   I recommend taking warm clothes for the winter.
You should avoid drinking tap water.

1. I recommend:

_____

_____

_____

2. You should avoid:

_____

_____

_____

EXERCISE **7**   ABOUT YOU Tell if you like or don't like the following activities. Explain why.

EXAMPLES   **go shopping**
I like to go shopping for clothes because I like to try new styles.
go bowling
I don't like to go bowling because I don't think it's an interesting sport.

1. go fishing   3. go jogging   5. go hunting
2. go camping   4. go swimming   6. go shopping

[3]*Do* makes the verb more emphatic. In this sentence, it shows contrast with *don't mind.*

10.4 | Gerund After Preposition[4]

A gerund can follow a preposition. It is important to choose the correct preposition after a verb or adjective.

Preposition combinations		Common combinations	Examples
Verb + Preposition	verb + *about*	care about complain about dream about forget about talk about think about worry about	I **care about doing** well on an interview. My sister **dreams about becoming** a doctor.
	verb + *to*	adjust to look forward to object to	I am **looking forward to getting** a job and **saving** money.
	verb + *on*	depend on insist on plan on	I **plan on going** to a career counselor.
	verb + *in*	believe in succeed in	My father **succeeded in finding** a good job.
Adjective + Preposition	adjective + *of*	afraid of capable of guilty of proud of tired of	I'm **afraid of losing** my job.
	adjective + *about*	concerned about excited about upset about worried about sad about	He is **upset about not getting** the job.
	adjective + *for*	responsible for famous for grateful to... for	Who is **responsible for hiring** in this company?
	adjective + *at*	good at successful at	I'm not very **good at writing** a résumé.
	adjective + *to*	accustomed to used to	I'm not **accustomed to talking** about my strengths.
	adjective + *in*	interested in successful in	Are you **interested in getting** a better job?

[4]For a list of verbs and adjectives followed by a *preposition*, see Appendix H.

Language Notes:

1. *Plan, afraid,* and *proud* can be followed by an infinitive too.

   I plan **on seeing** a counselor./ I plan **to see** a counselor.

   I'm afraid **of losing** my job./ I'm afraid **to lose** my job.

   He's proud **of being** a college graduate./He's proud **to be** a college graduate.

2. Notice that in some expressions, *to* is a preposition followed by a gerund, not part of an infinitive.

   Compare:

   I need **to write** a résumé. (infinitive)

   I'm not accustomed **to writing** a résumé. (*to* + gerund)

EXERCISE **8** ABOUT YOU Complete the questions with a gerund (phrase). Then ask another student these questions.

EXAMPLE Are you lazy about doing your homework? _____

1. Do you ever worry about _____
2. Do you plan on _____
3. Do you ever think about _____
4. When you get tired of _____, what do you do?
5. Are you interested in _____

EXERCISE **9** ABOUT YOU Fill in the blanks with a preposition and a gerund (phrase) to make a **true** statement.

EXAMPLE I plan on going back to Haiti soon. _____

1. I'm afraid _____
2. I'm not afraid _____
3. I'm interested _____
4. I'm not interested _____
5. I want to succeed _____
6. I'm not very good _____
7. I'm accustomed _____
8. I'm not accustomed _____
9. I plan _____
10. I don't care _____

EXERCISE **10**   ABOUT YOU Fill in the blanks to complete each statement. Compare your experiences in the U.S. with your experiences in your native country. You may share your answers with a small group or with the entire class.

EXAMPLES   In the U.S., I'm afraid of <u>walking alone at night.</u>
In my native country, I was afraid of <u>not being able to give my</u>
<u>children a good future.</u>

1. In the U.S., I'm interested in _____
   In my native country, I was interested in _____

   _____

2. In the U.S., I worry about _____
   In my native country, I worried about _____

   _____

3. In the U.S., I dream about _____
   In my native country, I dreamed about _____

   _____

4. In the U.S., I look forward to _____
   In my native country, I looked forward to _____

   _____

5. In the U.S., people often complain about _____
   In my native country, people often complain about _____

   _____

6. In the U.S., families often talk about _____
   In my native country, families often talk about _____

   _____

7. American students are accustomed to _____
   Students in my native country, are accustomed to _____

   _____

10.5 | Gerund in Adverbial Phrase

Examples	Explanation
You should practice interview questions **before going** on an interview. I found my job **by looking** in the newspaper. She took the test **without studying.**	We can use a gerund in an adverbial phrase that begins with a preposition: *before, by, after, without,* etc.

EXERCISE 11 Fill in the blanks to complete the sentences.

EXAMPLE The best way to improve your vocabulary is by _____reading._____

1. One way to find a job is by _____

2. It is very difficult to find a job without _____
   _____

3. The best way to improve your pronunciation is by _____
   _____

4. The best way to quit a bad habit is by _____

5. One way to find an apartment is by _____

6. I can't speak English without _____

7. It's impossible to get a driver's license without _____
   _____

8. You should read the instructions of a test before _____
   _____

EXERCISE 12 Fill in the blanks in the conversation below with the gerund form. Where you see two blanks, use a preposition before the gerund. Answers may vary.

A: I need to find a job. I've had ten interviews, but so far no job.

B: Have you thought _____about_____ _____going_____ to a job counselor?
   (example)

A: No. Where can I find one?

B: Our school office has a counseling department. I suggest _____
   _____ an appointment with a counselor.
   (1)

A: What can a job counselor do for me?

B: Do you know anything about interviewing skills?

A: No.

B: Well, with the job counselor, you can talk _____
(2)

_____ a good impression during an interview.
(3)

You can practice _____ questions that the
(4)

interviewer might ask you.

A: Really? How does the counselor know what questions the interviewer will ask me?

B: Many interviewers ask the same general questions. For example, the interviewer might ask you, "Do you enjoy _____
(5)

with computers?" Or she might ask you, "Do you mind _____

_____ overtime and on weekends?" Or "Are you good
(6)

_____   _____
(7)   (8)

with other people?"

A: I dislike _____ about myself.
(9)

B: That's what you have to do in the U.S.

A: What else can the counselor help me with?

B: If your skills are low, you can talk about _____
(10)

your skills. If you don't know much about computers, for example, she can recommend _____ more classes.
(11)

A: It feels like I'm never going to find a job. I'm tired _____
(12)

_____ and not finding anything.
(13)

B: If you keep _____, you will succeed
(14)

_____   _____ a job. I'm
(15)   (16)

sure. But it takes time and patience.

10.6 | Infinitives—An Overview

To form an infinitive, we use *to* + the base form of a verb (*to find, to help, to run, to be*).

Examples	Explanation
I want **to find** a job.	An infinitive is used after certain verbs.
I want you **to help** me.	An object can be added before an infinitive.
I'm happy **to help** you.	An infinitive can follow certain adjectives.
It's important **to write** a good résumé.	An infinitive follows certain expressions with *it*.
He went to a counselor **to get** advice.	An infinitive is used to show purpose.

TIPS[5] ON WRITING A RÉSUMÉ

Before You Read

1. Have you ever written a résumé? What is the hardest part about writing a résumé?
2. Do people in your native country have to write a résumé?

 Read the following article. Pay special attention to infinitives.

It's important **to write** a good, clear résumé. A résumé should be limited to one page. It is only necessary **to describe** your most relevant work.[6] Employers are busy people. Don't expect them **to read** long résumés.

You need **to present** your abilities in your résumé. Employers expect you **to use** action verbs **to describe** your experience. Don't begin your sentences with "I". Use past tense verbs like: *managed, designed, created,* and *developed*. It is not enough **to say** you improved something. Be specific. How did you improve it?

Before making copies of your résumé, it is important **to check** the grammar and spelling. Employers want **to see** if you have good communications skills. Ask a friend or teacher **to read** and **give** an opinion about your résumé.

It isn't necessary **to include** references. If the employer wants you **to provide** references, he or she will ask you **to do** so during or after the interview.

Don't include personal information such as marital status, age, race, family information, or hobbies.

Be honest in your résumé. Employers can check your information. No one wants **to hire** a liar.

---

[5]A *tip* is a small piece of advice.

[6]*Relevant work* is work that is related to this particular job opening.

TINA WHITE

1234 Anderson Avenue

West City, MA 01766

tina.white@met.com

617-123-1234 (home)

617-987-9876 (cellular)

EXPERIENCE

COMPUTER SALES MANAGER

Acme Computer Services, Inc., Concord, MA

March 2003-Present

- Manage computer services department, overseeing 20 sales representatives throughout New England.
- Exceeded annual sales goal by 20 percent in 2004.
- Created online customer database, enabling representatives and company to track and retain customers and improve service.
- Developed new training program and materials for all company sales representatives.

OFFICE MANAGER

West Marketing Services, West City, MA

June 1999-March 2003

- Implemented new system for improving accounting records and reports.
- Managed, trained, and oversaw five customer service representatives.
- Grew sales contracts for support services by 200 percent in first two years.

EDUCATION AND TRAINING

Northeastern Community College, Salem, MA

  Associates Degree Major: Accounting

Institute of Management, Boston, MA

  Certificate of Completion. Course: Sales Management

COMPUTER SKILLS

Proficient in use of MS Windows, PowerPoint, Excel, Access,

  Outlook, MAC OS, and several accounting and database systems

10.7 | Infinitive as Subject

An infinitive can be the subject of a sentence. We begin the sentence with it and delay the infinitive.

Examples	Explanation
It is important **to write** a good résumé. It isn't necessary **to include** all your experience. It takes time **to find** a job.	We can use an infinitive after these adjectives: dangerous good necessary difficult great possible easy hard sad expensive important wrong fun impossible
It is necessary **for the manager** to choose the best candidate for the job. It isn't easy **for me** to talk about myself. It was hard **for her** to leave her last job.	Include *for* + noun or object pronoun to make a statement that is true of a specific person.
Compare Infinitive and Gerund Subjects: It's important **to arrive** on time. **Arriving** on time is important.	There is no difference in meaning between an infinitive subject and a gerund subject.

EXERCISE 13   Fill in the blanks with an appropriate infinitive to give information about résumés and interviews. Answers may vary.

EXAMPLE   It is necessary _____ to have _____ a Social Security card.

1. It isn't necessary _____ *all* your previous experience. Choose only the most relevant experience.

2. It's important _____ your spelling and grammar before sending a résumé.

3. It is a good idea _____ interview questions before going on an interview.

4. It is important _____ your best when you go on an interview, so choose your clothes carefully.

5. It isn't necessary _____ references on a résumé. You can simply write, "References available upon request."

6. It's important _____ your past work experience in detail, using words like *managed*, *designed*, *supervised*, and *built*.

EXERCISE 14 Complete each statement with an infinitive phrase. You can add an object, if you like.

EXAMPLES It's easy _to shop in an American supermarket._
It's necessary _for me to pay my rent by the fifth of the month._

1. It's important _____
2. It's impossible _____
3. It's possible _____
4. It's necessary _____
5. It's dangerous _____
6. It isn't good _____
7. It's expensive _____
8. It's hard _____

EXERCISE 15 ABOUT YOU Tell if it's important or not important for you to do the following.

EXAMPLE own a house
It's (not) important for me to own a house.

1. get a college degree
2. find an interesting job
3. have a car
4. speak English well
5. read and write English well

6. study American history
7. become an American citizen
8. own a computer
9. have a cell phone
10. make a lot of money

EXERCISE 16 Write a sentence with each pair of words below. You may read your sentences to the class.

EXAMPLE hard / the teacher

It's hard for the teacher to pronounce the names of some students.

1. important / us (the students)

_____

2. difficult / Americans

_____

3. easy / the teacher

_____

4. necessary / children

_____

5. difficult / a woman

_____

6. difficult / a man

_____

EXERCISE **17** Write a list of things that a foreign student or immigrant should know about life in the U.S. Use gerunds or infinitives as subjects. You may work with a partner.

EXAMPLES It is possible for some students to get financial aid.

Learning English is going to take longer than you expected.

1. _____
2. _____
3. _____
4. _____
5. _____
6. _____

10.8 | Infinitive After Adjective

Some adjectives can be followed by an infinitive.	
Examples	Explanation
I would be happy **to help** you with your résumé. Are you prepared **to make** copies of your résumé?	Adjectives often followed by an infinitive are: afraid happy prepared ready glad lucky proud sad

EXERCISE **18** Complete this conversation with appropriate infinitives. Answers may vary.

EXAMPLE A: I have my first interview tomorrow. I'm afraid ____to go____

(example)

alone. Would you go with me?

B: I'd be happy _____ with you and wait in the car. But

(1)

nobody can go with you on an interview. You have to do it alone. It

sounds like you're not ready _____ a job interview. You

(2)

should see a job counselor and get some practice before you have

an interview. I was lucky _____ a great job counselor. She

(3)

prepared me well.

A: I don't have time to make an appointment with a job counselor before

tomorrow. Maybe you can help me.

B: I'd be happy _____ you. Do you have some time this

(4)

afternoon? We can go over some basic questions.

A: Thanks. I'm glad _____ you as my friend.

(5)

B: That's what friends are for.

EXERCISE **19** ABOUT YOU Fill in the blanks.

EXAMPLE I'm lucky _to be in the U.S._

1. I was lucky _____

2. I'm proud _____

3. I'm sometimes afraid _____ alone.

4. I'm not afraid _____

5. In the U.S., I'm afraid _____

6. Are we ready _____

7. I'm not prepared _____

10.9 | Infinitive After Verb

	Some verbs are commonly followed by an infinitive (phrase).	

Examples	Explanation
I need **to find** a new job.	We can use an infinitive after the following
I decided **to quit** my old job.	verbs:
I prefer **to work** outdoors.	agree decide like promise
I want **to make** more money.	ask expect love refuse
	attempt forget need remember
	begin hope plan start
	continue learn prefer try

Pronunciation Note:

The *to* in infinitives is often pronounced "ta" or, after a *d* sound, "da." *Want* to is often pronounced "wanna." Listen to your teacher pronounce the sentences in the above box.

EXERCISE 20   ABOUT YOU Ask a question with the words given in the present tense. Another student will answer.

EXAMPLE   like / work with computers

A: Do you like to work with computers?
B: Yes, I do.   OR   No, I don't.

1. plan / look for a job
2. expect / make a lot of money at your next job
3. like / work with computers
4. prefer / work the second shift
5. need / see a job counselor
6. hope / become rich some day
7. like / work with people
8. try / keep up with changes in technology
9. want / learn another language
10. continue / speak your native language at home

EXERCISE **21**   ABOUT YOU Write a sentence about yourself, using the words given, in any tense. You may share your sentences with the class.

EXAMPLES   **like / eat**

I like to eat Chinese food.

try / find

I'm trying to find a job.

1. **like / read**

_____

2. **not like / eat**

_____

3. **want / visit**

_____

4. **decide / go**

_____

5. **try / learn**

_____

6. **begin / study**

_____

EXERCISE **22**   ABOUT YOU Check (√) the activities that you like to do. Tell the class why you like or don't like this activity.

1. _____ stay home on the weekends   5. _____ go to museums
2. _____ eat in a restaurant   6. _____ dance
3. _____ get up early   7. _____ write letters
4. _____ talk on the phone   8. _____ play chess

chess

10.10 | Gerund or Infinitive After Verb

Some verbs can be followed by either a gerund or an infinitive with almost no difference in meaning.

Examples	Explanation
I started **looking** for a job a month ago. I started **to look** for a job a month ago. He continued **working** until he was 65 years old. He continued **to work** until he was 65 years old.	The verbs below can be followed by either a gerund or an infinitive with almost no difference in meaning: attempt deserve prefer begin hate start can't stand[7] like try continue love

Language Notes:

1. The meaning of *try* + infinitive is a little different from the meaning of *try* + gerund.
   Try + infinitive means to make an effort.
   > I'll **try to improve** my résumé.
   > You should **try to relax** during the interview.

2. *Try* + gerund means to use a different technique when one technique doesn't produce the result you want.
   > I wanted to reach you yesterday, but I couldn't. I **tried calling** your home phone, but I got your answering machine. I **tried calling** your cell phone, but it was turned off. I **tried e-mailing** you, but you didn't check your e-mail.

EXERCISE 23   ABOUT YOU Complete each statement using either a gerund (phrase) or an infinitive (phrase). Practice both ways.

EXAMPLES   I started to learn English four years ago.
                          (learn)

   I started studying French when I was in high school.
                        (study)

1. I started _____ to this school in _____
                  (come)

2. I began _____ English _____
             (study)

3. I like _____ on TV.
                  (watch)

4. I like _____
                  (live)

5. I hate _____
                  (wear)

6. I love _____
                  (eat)

[7] *Can't stand* means hate or can't tolerate. I *can't stand* waiting in a long line.

10.11 | Object Before Infinitive

We can use a noun or object pronoun (*me, you, him, her, it, us,* and *them*) before an infinitive.

Examples	Explanation
Don't **expect an employer to read** a long résumé. I **want you to look** at my résumé. My boss **wants me to work** overtime. I **expected him to give** me a raise.	We often use an object between the following verbs and an infinitive: advise invite want allow need would like ask permit expect tell
He helped me **find** a job. He helped me *to* **find** a job	*Help* can be followed by either an object+base form or an object+infinitive.

EXERCISE 24 Fill in the blanks with pronouns and infinitives to complete the conversation below.

A: I want to quit my job.

B: Why?

A: I don't like my supervisor. He expects ____me____ ____to work____

                                           (example)   *(example: work)*

at night and on weekends.

B: But you get extra pay for that, don't you?

A: No. I asked _____ _____ me a raise,

                             (1)   *(2 give)*

but he said the company can't afford it.

B: Is that the only problem?

A: No. My co-workers and I like to go out for lunch. But he doesn't

want _____ _____ out. He

              (3)   *(4 go)*

expects _____ _____ in the company

              (5)   *(6 eat)*

cafeteria. He says that if we go out, we might not get back on time.

B: That's awful. He should permit _____

                                                  (7)

_____ wherever you want to.

     (8 eat)

A: That's what I think. I also have a problem with my manager. She never gives anyone a compliment. When I do a good job, I expect _____ _____something nice. But she
            (9)   *(10 say)*

only says something when we make a mistake.

B: It's important to get positive feedback too.

A: Do you know of any jobs in your company? I'd like _____
_____ _____ your boss if he
        (11)   *(12 ask)*

needs anyone.

B: I don't think there are any job openings in my company. My boss has two sons in their twenties. He wants _____
                                     (13)

_____ for him on Saturdays. But they're so
       (14 work)

lazy. The boss allows _____ _____
                       (15)   *(16 come)*

late and _____ early. He would never permit
       (17 leave)

_____ _____that. We have to be
     (18)   *(19 do)*

on time exactly, or he'll take away some of our pay.

A: Maybe I should just stay at my job. I guess no job is perfect.

EXERCISE **25** Tell if the teacher wants or doesn't want the students to do the following.

EXAMPLES   do the homework
The teacher wants us to do the homework.

use the textbook during a test
The teacher doesn't want us to use the textbook during a test.

1. talk to another student during a test
2. study before a test
3. copy another student's homework in class
4. learn English
5. speak our native languages
6. improve our pronunciation

EXERCISE 26   ABOUT YOU Tell if you expect or don't expect the teacher to do the following.

EXAMPLES   give homework
I expect him / her to give homework.
give me private lessons
I don't expect him / her to give me private lessons.

1. correct the homework   6. pass all the students
2. give tests   7. know a lot about my native country
3. speak my native language   8. answer my questions in class
4. help me after class   9. teach us American history
5. come to class on time   10. pronounce my name correctly

EXERCISE 27   ABOUT YOU Write sentences to tell what one member of your family wants (or doesn't want) from another member of your family.

EXAMPLES   My father doesn't want my brother to watch so much TV.
My brother wants me to help him with his math homework.

1. _____
2. _____
3. _____
4. _____

10.12 | Infinitive to Show Purpose

We use the infinitive to show purpose.

Examples	Explanation
You can use the Internet **to find** job information. I need a car **to get** to work. I'm saving my money **to buy** a car.	*To* is the short form of *in order to*.
You can use the Internet **in order to find** job information. I need a car **in order to get** to work. I'm saving my money **in order to buy** a car.	The long form is *in order to*.

EXERCISE 28   Fill in the blanks with an infinitive to show purpose. Answers will vary.

EXAMPLE   I bought the Sunday newspaper _____ to look for a job _____.

1. I called the company _____ an appointment.
2. She wants to work overtime _____

3. You should use the Internet _____ jobs.

4. You can use a résumé writing service _____
   your résumé.

5. My interview is in a far suburb. I need a car _____
   the interview.

6. Use express mail _____ faster.

7. In the U.S., you need experience _____,
   and you need a job _____.

8. I need two phone lines. I need one _____
   on the phone with my friends and relatives. I need the other one
   _____ business calls.

9. I'm sending a letter that has a lot of papers in it. I need extra
   stamps _____ this letter.

10. You should go to the college admissions office _____
    _____ a copy of your transcripts.

11. After an interview, you can call the employer _____
    _____ that you're very interested in the position.

RITA'S STORY

Before You Read

1. What are some differences in the American workplace and the workplace in other countries?

2. In your native culture, is it a sign of respect or disrespect to look at someone directly?

Read the following story. Pay special attention to *used to*, *be used to*, and *get used to*.

I've been in the U.S. for two years. I **used to** study British English in India, so I had a hard time **getting used to** the American pronunciation. But little by little, I started to **get used to** it. Now I understand Americans well, and they understand me.

I **used to be** an elementary school teacher in India. But for the past two years in the U.S, I've been working in a hotel cleaning rooms. I have to work the second shift. I'm **not used to working** nights. I don't like it because I don't see my children very much. When I get home from work, they're asleep.

(continued)

My husband is home in the evening and cooks for them. In India, I **used to do** all the cooking, but now he has to help with household duties. He didn't like it at first, but now he's **used to it.**

When I started looking for a job, I had to **get used to** a lot of new things. For example, I had to learn to talk about my abilities in an interview. In India, it is considered impolite to say how wonderful you are. But my job counselor told me that I had to **get used to** it because that's what Americans do. Another thing I'**m not used** to is wearing American clothes. In India, I **used to wear** traditional Indian clothes to work. But now I wear a uniform to work. I don't like to dress like this. I prefer traditional Indian clothes, but my job requires a uniform. There's one thing I **can't get used to:** everyone here calls each other by their first names. It's our native custom to use a term of respect with people we don't know.

It has been hard to **get used to** so many new things, but little by little, I'm doing it.

10.13 | *Used To* vs. *Be Used To*

Used to + base form is different from *be used* to + gerund.	

Examples	Explanation
Rita **used to be** an elementary school teacher. Now she cleans hotel rooms.	*Used to* + base form tells about a past habit or custom. This activity has been discontinued.
She **used to wear** traditional Indian clothes. Now she wears a uniform to work.	
She **used to cook** dinner for her family in India. Now her husband cooks dinner.	
Her husband **didn't use to cook** in India.	The negative is *didn't use to* + *base* form. (Remove the *d* at the end.)
I**'m used to working** in the day, not at night. Women in India **are used to wearing** traditional clothes.	*Be used to* + gerund or noun means *be accustomed to*. Something is a person's custom and is therefore not difficult to do.
People who studied British English **aren't used to the American pronunciation.**	The negative is *be* + *not* + *used to* + gerund or noun. (Do not remove the *d* of **used to**.)
If you emigrate to the U.S., you have to **get used to many new things.**	*Get used to* + gerund or noun means *become accustomed to*.
Children from another country usually **get used to living** in the U.S. easily. But it takes their parents a long time to **get used to a new life.**	
I **can't get used to** the cold winters here. She **can't get used to** calling people by their first names.	For the negative, we usually say *can't get used to*.

EXERCISE 29   ABOUT YOU Write four sentences comparing your former behaviors to your behaviors or customs now.

EXAMPLES   I used to live with my family. Now I live with a roommate.

I used to worry a lot. Now I take it easy most of the time.

1. _____
2. _____
3. _____
4. _____

EXERCISE 30   ABOUT YOU Write sentences comparing the way you used to live in your country or in another city and the way you live now. Read your sentences to the class.

EXAMPLE   I used to go everywhere by bus. Now I have a car.

1. _____

2. _____

3. _____

4. _____

EXERCISE 31   A student wrote about things that are new for her in an American classroom. Fill in the blanks with a gerund. Then tell if *you* are used to these things or not.

EXAMPLE   I'm not used to _____taking_____ multiple-choice tests. In my native country, we have essay tests.

1. I'm not used to _____ at small desks. In my native country, we sit at large tables.

2. I'm not used to _____ the teacher by his / her first name. In my country, we say "Professor."

3. I'm not used to _____ in a textbook. In my native country, we don't write in the books because we borrow them from the school.

4. I'm not used to _____ jeans to class. In my native country, students wear a uniform.

5. I'm not used to _____ and studying at the same time. Students in my native country don't work. Their parents support them.

6. I'm not used to _____ a lot of money to attend college. In my native country, college is free.

7. I'm not used to _____ when a teacher asks me a question. In my native country, students stand to answer a question.

EXERCISE 32 ABOUT YOU Name four things that you had to get used to in the U.S. or in a new town or school. (These things were strange for you when you arrived.)

EXAMPLES I had to get used to living in a small apartment. _____
I had to get used to American pronunciation. _____

1. I had to get used to _____
2. I had to get used to _____
3. I had to get used to _____
4. I had to get used to _____

EXERCISE 33 ABOUT YOU Answer each question with a complete sentence. Practice *be used to* + gerund or noun.

EXAMPLE What are you used to drinking in the morning?
I'm used to drinking coffee in the morning.

1. What kind of work are you used to?
2. What kind of relationship are you used to having with co-workers?
3. What kind of food are you used to (eating)?
4. What kind of weather are you used to?
5. What time are you used to getting up?
6. What kinds of clothes are you used to wearing to work or class?
7. What kinds of things are you used to doing every day?
8. What kinds of classroom behaviors are you used to?
9. What kinds of things are you used to doing alone?

EXERCISE **34**   Circle the correct words to complete this conversation.

A: How's your new job?

B: I don't like it at all. I have to work the night shift. I can't get used to
   (sleep / sleeping) during the day.
      (1)

A: I know. That's hard. I used to *(work / working)* the night shift, and I
                                      (2)
   hated it. That's why I quit.

B: But the night shift pays more money.

A: I know it does, but I was never home for my children. Now my kids
   speak more English than Spanish. They used to *(speaking / speak)*
                                                      (3)
   Spanish well, but now they mix Spanish and English. They play with
   their American friends all day or watch TV.

A: My kids are the same way. But *(I'm / I)* used to it. It doesn't bother me.
                                    (4)

B: I can't *(get / be)* used to it. My parents came to live with us, and they
              (5)
   don't speak much English. So they can't communicate with their
   grandchildren anymore.

A: My parents used to *(living / live)* with us too. But they went back
                          (6)
   to Mexico. They didn't like the winters here. They couldn't get
   (use / used) to the cold weather.
      (7)

A: Do you think Americans are *(used to / use to)* cold weather?
                                  (8)

B: I'm not sure. My coworker was born in the U.S., but she says she
   hates winter. She *(is used to / used to)* live in Texas, but now she
                          (9)
   lives here in Minnesota.

A: Why did she move here if she hates the cold weather?

B: The company where she used to *(work / working)* closed down and
                                      (10)
   she had to find another job. Her cousin helped her find a job here.

A: Before I came to the U.S., I thought everything here would be perfect.
   I didn't *(use / used)* to *(think / thinking)* about the
              (11)   *(12)*
   problems. But I guess life in every country has its problems.

SUMMARY OF LESSON 10

Gerunds	
Examples	Explanation
Working all day is hard.	As the subject of the sentence
I don't enjoy **working** as a taxi driver.	After certain verbs
I **go shopping** after work.	In many idiomatic expressions with *go*
I'm worried about **finding** a job.	After prepositions
She found a job by **looking** in the newspaper.	In adverbial phrases

Infinitives	
Examples	Explanation
I need **to find** a new job.	After certain verbs
My boss wants me **to work** overtime.	After an object
I'm ready **to quit**.	After certain adjectives
It's important **to have** some free time. It's impossible for me **to work** 80 hours a week.	After certain impersonal expressions beginning with *it.*
I work (in order) **to support** my family.	To show purpose

Gerund or Infinitive—No Difference in Meaning	
Gerund	Infinitive
I like **working** with computers.	I like **to work** with computers.
I began **working** three months ago.	I began **to work** three months ago.
Writing a good résumé is important.	It's important **to write** a good résumé.

Gerund or Infinitive—Difference in Meaning	
Infinitive (Past Habit)	Gerund (Custom)
Rita **used to be** a teacher in India. Now she works in a hotel.	She **isn't used to working** the night shift. It's hard for her.
Rita **used to wear** traditional Indian clothes to work. Now she wears a uniform.	Rita studied British English. She had to **get used to hearing** the American pronunciation.

1. Use a gerund after a preposition.

   using
   He read the whole book without ~~use~~ a dictionary.

2. Use the correct preposition.

   on
   She insisted ~~in~~ driving me home.

3. Use a gerund after certain verbs.

   ing
   I enjoy ~~to~~ walk_∧ in the park.

   ping
   He went ~~to~~ shop_∧ after work.

4. Use an infinitive after certain verbs.

   to
   I decided_∧ buy a new car.

5. Use a gerund, not a base form, as a subject.

   Finding
   ~~Find~~ a good job is important.

6. Don't forget to include *it* for a delayed infinitive subject.

   It i
   ~~Is~~ important to find a good job.

7. Don't use the past form after to.

   buy
   I decided to ~~bought~~ a new car.

8. After *want, expect, need, advise,* and *ask,* use an object pronoun, not a subject pronoun, before the infinitive. Don't use *that* as a connector.

   me to
   He wants ~~that I~~ drive.

   us to
   The teacher expects ~~we~~ do the homework.

9. Use *for,* not *to,* when introducing an object after impersonal expressions beginning with *it.* Use the object pronoun after *for.*

   for
   It's important ~~to~~ me to find a job.

   him
   It's necessary for ~~he~~ to be on time.

10. Use *to* + base form, not *for*, to show purpose.

> I called the company ~~for~~ to make an appointment.

11. Don't put *be* before *used to* for the habitual past.

> ~~I'm~~ used to live in Germany. Now I live in the U.S.

12. Don't use the *-ing* form after *used to* for the habitual past.

> We used to ~~having~~ have a dog, but he died.

13. Don't forget the *d* in *used to*.

> I use d to live with my parents. Now I live alone.

LESSON 10 TEST/REVIEW

PART **1** Find the mistakes with the underlined words, and correct them. Not every sentence has a mistake. If the sentence is correct, write *C*.

EXAMPLES He wrote the composition without ~~check~~ checking his spelling.

Do you <u>like to play</u> tennis?   C

1. <u>Using</u> the Internet is fun.
2. I recommend <u>to see</u> a job counselor.
3. Do you enjoy <u>learning</u> a new language?
4. <u>Save</u> your money is important for your future.
5. It's important <u>to me</u> to know English well.
6. It's impossible <u>for she</u> to work 60 hours a week.
7. Do you <u>go fish</u> with your brother every week?
8. Do you <u>want the teacher to review</u> modals?
9. He got rich <u>by working</u> hard and <u>investing</u> his money.
10. The teacher <u>tried to explained</u> the present perfect, but we didn't understand it.
11. <u>Is</u> necessary to come to class on time.
12. Do you <u>want to watch</u> TV?
13. She <u>use to take</u> the bus every day. Now she has a car and drives everywhere.
14. It's necessary <u>for me have</u> a good education.

15. She came to the U.S. <u>for find</u> a better job.

16. He's interested <u>in becoming</u> a nurse.

17. We're thinking <u>to spend</u> our vacation in Acapulco.

18. We've always lived in a big city. <u>We're used to living</u> in a big city.

19. <u>I'm used to live</u> with my family. Now I live alone.

20. She's from England. She can't <u>get used to drive</u> on the right side of the road.

21. My mother <u>wants that I call</u> her every day.

22. Are you worried about <u>lose</u> your job?

PART **2** Fill in the blanks in the conversation below. Use a gerund or an infinitive. In some cases, either the gerund or the infinitive is possible. Answers may vary.

A: Hi, Molly. I haven't seen you in ages. What's going on in your life?

B: I've made many changes. First, I quit ____working____ in a factory.
(example)

   I disliked _____ the same thing every day. And I
   (1)

   wasn't used to _____ on my feet all day. My boss
   (2)

   often wanted me _____ overtime on Saturdays. I
   (3)

   need _____ with my children on Saturdays.
   (4)

   Sometimes they want me _____ them to the
   (5)

   zoo or to the museum. And I need _____ them
   (6)

   with their homework too.

A: So what do you plan on _____ ?
   (7)

B: I've started _____ to college
   (8)

   _____ some general courses.
   (9)

A: What career are you planning?

B: I'm not sure. I'm interested in _____ with children.
   (10)

   Maybe I'll become a teacher's aide. I've also thought about

   _____ in a day care center. I care about
   (11)

   _____ people.
   (12)

348   Lesson 10

A: Yes, it's wonderful _____ other people, especially

    (13)

children. It's important _____ a job that you like.

    (14)

So you're starting a whole new career.

B: It's not new, really. Before I came to the U.S., I used _____

_____ a kindergarten teacher in my country. But

    (15)

my English wasn't so good when I came here, so I found a job in a

factory. I look forward to _____ to my former

    (16)

profession or doing something similar.

A: How did you learn English so fast?

B: By _____ with people at work, by

    (17)

_____ TV, and by _____ the

    (18)   (19)

newspaper. It hasn't been easy for me _____

    (20)

American English. I studied British English in my country, but here I

have to get used to _____ things like "gonna" and

    (21)

"wanna." At first I didn't understand Americans, but now I'm used

to their pronunciation. I've had to make a lot of changes.

A: You should be proud of _____ so many changes

    (22)

in your life so quickly.

B: I am.

A: Let's get together some time and talk some more.

B: I'd love to. I love to dance. Maybe we can go _____

    (23)

together sometime.

A: That would be great. And I love _____ .

    (24)

Maybe we can go shopping together sometime.

Classroom Activities

1. If you have a job, write a list of five things you enjoy and don't enjoy about your job. If you don't have a job, you can write about what you enjoy and don't enjoy about this school or class. Share your answers with the class.

I enjoy:	I don't enjoy:
I enjoy talking to people.	I don't enjoy working at 6 a.m.

2. Compare the work environment in the U.S. to the work environment in another country. Discuss your answers in a small group or with the entire class. (If you have no experience with American jobs, ask an American to fill in his / her opinions about the U.S.)

	The U.S.	Another Country
1. Coworkers are friendly with each other at the job.		
2. Coworkers get together after work to socialize.		
3. Arriving on time for the job is very important.		
4. The boss is friendly with the employees.		
5. The employees are very serious about their jobs.		
6. The employees use the telephone for personal use.		
7. Everyone wears formal clothes.		
8. Employees get long lunch breaks.		
9. Employees get long vacations.		
10. Employees call the company if they are sick and can't work on a particular day.		
11. Employees are paid in cash.		
12. Employees often take work home.		

3. Find a partner. Pretend that one of you is the manager of a company and the other one is looking for a job in that company. First decide what kind of company it is. Then write the manager's questions and the applicant's answers. Perform your interview in front of the class.

Talk About it

1. Talk about your experiences in looking for a job in the U.S.
2. Talk about the environment where you work.
3. Talk about some professions that interest you.
4. Talk about some professions that you think are terrible.

Write About it

1. Write your résumé and a cover letter.
2. Write about a job you wouldn't want to have. Tell why.
3. Write about a profession you would like to have. Tell why.
4. Write about your current job or a job you had in the past. Tell what you like(d) or don't (didn't) like about this job.

Internet Activities

1. Type *career* in a search engine. See how many "hits" come up.
2. Find some career counseling Web sites. Find a sample résumé in your field or close to your field. Print it out and bring it to class.
3. From one of the Web sites you found, get information on one or more of the following topics:

   - how to write a cover letter
   - how to find a career counselor
   - how to plan for your interview
   - how to network
   - what questions to ask an interviewer

4. See if your local newspaper has a Web site. If it does, find the Help Wanted section of this newspaper. Bring job listings that interest you to class.

Additional Activities at http://elt.thomson.com/gic

GRAMMAR

Adjective Clauses

CONTEXT: Making Connections—Old Friends and New

Finding Old Friends

Internet Matchmaking

11.1 | Adjective Clauses—An Overview

An adjective is a word that describes a noun.   An adjective clause is a group of words (with a subject and a verb) that describes a noun. Compare **adjectives** (ADJ.) and adjective clauses (AC) below.

Examples	Explanation
ADJ: Do you know your **new** neighbors? **AC:** Do you know the people **who live next door to you?** **ADJ:** This is an **interesting** book. **AC:** This is a book **that has pictures of the high school graduates.** **ADJ:** I attended an **old** high school. **AC:** The high school **that I attended** was built in 1920.	An adjective (ADJ) precedes a noun. An adjective clause (AC) follows a noun. Relative pronouns, such as *who* and *that*, introduce an adjective clause.

FINDING OLD FRIENDS

Before You Read

1. Do you keep in touch with old friends from elementary school or high school?

2. Have you ever thought about contacting someone you haven't seen in years?

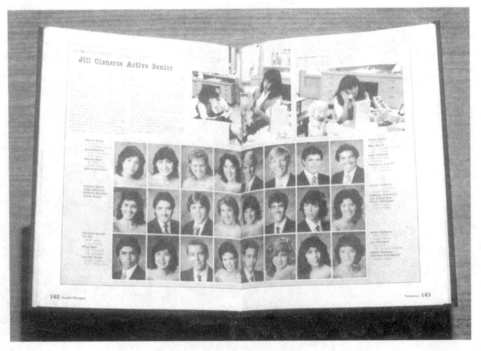

High School Yearbook

Americans move numerous times during their lives. As a result, they often lose touch with old friends. Usually, during their twenties and thirties, people are too busy building their careers and starting their families to think much about the past. But as people get older, they often start to wonder about the best friend **they had in high school,** the soldier **with whom they served in the military,** the person **who lived next door** when they were growing up, or their high school sweetheart. Many people want to connect with the past.

Before the Internet, finding a lost love or an old friend required searching through old phone books in libraries in different cities, a detective, and a lot of luck. It was especially hard to find married women **who changed their names.**

Now with the Internet, old friends can sometimes find each other in seconds. Several Web sites have emerged to meet people's growing desire to make connections with former classmates. There are Web sites **that list the students in high schools and colleges in the U.S.** People **who went to high school in the U.S.** can list themselves according to the school **they attended** and the year **they graduated.** A man might go to these Web sites looking for the guys **he played football with** or a long-lost friend—and find the name of a first love **whom he hasn't seen in years.**

One Web site, Classmates.com, claims that more than 30 million Americans have listed themselves on their site. Married women **who have changed their names** list themselves by their maiden names so that others can recognize them easily.

Another way **that people make connections with old classmates** is through reunions. Some high school graduating classes meet every ten years. They usually have dinner, remember the time **when they were young,** and exchange information about what they are doing today. They sometimes bring their high school yearbooks, **which have pictures of the graduates** and other school memories.

Some classes have their reunions in the schools **where they first met.** Others have their reunions in a nice restaurant. There are Web sites **that specialize in helping people find their former classmates and plan reunions.**

In America's highly mobile society, it takes some effort to connect with old friends. Looking back at fond memories, renewing old friendships, making new friends, and even starting a new romance with an old love can be the reward for a little work on the Internet.

EXERCISE **1**   Tell if the statement is true or false based on the reading on page 355. Write *T* or *F*.

EXAMPLE   People who graduate from high school have to attend their reunions. F

1. A yearbook is a book that has the diplomas of the graduates.
2. Classmates.com is a Web site that has lists of graduates from various high schools in the U.S.
3. Americans move a lot and often lose touch with the friends that they had in high school.
4. Women who get married often change their last names.
5. People who attend reunions meet their old classmates.
6. There are several Web sites that help people make connections with old friends.
7. Some Web sites can help you find people with whom you served in the military.

EXERCISE **2**   Underline the adjective clauses in the sentences in Exercise 1. Not every sentence has an adjective clause.

EXAMPLE   People who graduate from high school have to attend their reunions.

11.2 | Relative Pronoun as Subject

The relative pronouns *who*, *that*, and *which* can be the subject of the adjective clause. Use *who* or *that* for people. Use *that* or *which* for things.

Subject

I found a Web site. *The Web site* lists people by high school.

I found a Web site | that / which | lists people by high school.

Women .... often change their last names.

Subject

Women get married.

Women | who / that | get married often change their last names.

Language Notes:
1. *Which* is less common than *that*.
2. A present tense verb in the adjective clause must agree in number with its subject.
   A woman who **gets** married usually changes her name.
   Women who **get** married usually change their names.

EXERCISE **3**   Fill in the blanks with *who or that*+the correct form of the verb in parentheses ( ).

EXAMPLE   A yearbook has photos ___that show___ the activities of the high school.
                                    (show)

1. He has a yearbook _____ pictures of all his classmates.
                        (have)

2. People _____ to a reunion exchange information
            (go)
   about their lives.

3. Classmates.com is a Web site _____ people make
                                    (help)
   connections with old friends.

4. There are Web sites _____ in helping people
                          (specialize)
   plan a reunion.

5. People _____ a reunion have to contact former classmates.
            (plan)

EXERCISE **4**   Fill in the blanks with *who* or *that* + the correct form of the verb in parentheses ( ). Then complete the statement. Answers will vary.

EXAMPLE   People ___who work___ hard ___are often successful.___
                    (work)

1. People _____ regularly _____
              (exercise)

2. A person _____ a cell phone while driving _____
              (use)

   _____

3. Students _____ absent a lot _____
              (be)

4. Schools _____ computers _____
            (not/have)

5. A computer _____ more than five years old _____
                  (be)

   _____

6. People _____ digital cameras _____
            (have)

7. Colleges _____ evening classes _____
              (have)

8. A college _____ a day-care center _____
              (have)

   _____

9. Students _____ a full-time job _____
              (have)

EXERCISE 5 Complete each statement with an adjective clause. Answers will vary.

EXAMPLE I know some women _who don't want to get married._

1. People _____ can make a lot of friends.
2. Men _____ have a busy social life.
3. I like people _____
4. I don't like people _____
5. Students like a teacher _____
6. People _____ are very fortunate.
7. People _____ aren't usually successful.
8. Parents _____ are good.
9. A college _____ is good for foreign students.
10. People _____ have a hard life.

11.3 | Relative Pronoun as Object

The relative pronouns *who (m)*, *that*, and *which* can be the object of the adjective clause.

	Object
	She attended *the high school.*
The high school .... is in New York City.	
The high school [which / that / Ø] she attended is in New York City.	

	Object
	I knew a *friend* in high school.
A friend .... sent me an e-mail.	
A friend [who(m) / that / Ø] I knew in high school sent me an e-mail.	

Language Notes:

1. The relative pronoun is usually omitted in conversation when it is the object of the adjective clause.

    The high school she attended is in New York City.

2. *Whom* is considered more correct or more formal than *who* when used as the object of the adjective clause. However, as seen in the above note, the relative pronoun is usually omitted altogether in conversation.

    Formal: A friend *whom* I knew in high school sent me an e-mail.

    Informal: A friend I knew in high school sent me an e-mail.

EXERCISE **6**   In each sentence below, underline the adjective clause.

EXAMPLE   I've lost touch with some of the friends <u>I had in high school</u>.

1. The high school I attended is in another city.
2. The teachers I had in high school are all old now.
3. We didn't have to buy the textbooks we used in high school.
4. She married a man she met at her high school reunion.
5. The friends I've made in this country don't know much about my country.

EXERCISE **7**

A mother (M) is talking to her teenage daughter (D). Fill in the blanks to complete the conversation. Answers may vary.

M: I'd like to contact an old friend

I ____had____ in high school.
      (example)

I wish I could find her. I'll never

forget the good times _____ in high school. When
                              (1)

we graduated, we said we'd always stay in touch. But then we went to different colleges.

D: Didn't you keep in touch by e-mail?

M: When I was in college, e-mail didn't exist. At first we wrote letters. But little by little we wrote less and less until, eventually, we stopped writing.

D: Do you still have the letters she _____?
                                              (2)

M: Yes, I do. They're in a box in the basement.

D: Why don't you write to the address on the letters?

M: That doesn't make sense. The address she _____
                                                      (3)

on the letters was of the college town where she lived. I don't know what happened to her after she left college.

D: Have you tried calling her parents?

M: The phone number _____ is now
                              (4)

disconnected. Maybe her parents have died.

D: Have you looked on Classmates.com?

M: What's that?

D: It's a Web site that _____ lists of people. The list is
(5)

categorized by the high school you _____ and the
(6)

dates you _____ there.
(7)

M: Is everyone in my high school class on the list?

D: Unfortunately, no. Only the people _____
(8)

add their names are on the list.

M: But my friend probably got married. I don't know the name of the

man _____ married.
(9)

D: That's not a problem. You can search for her by her maiden name.

M: Will this Web site give me her address and phone number?

D: No. But for a fee, you can send her an e-mail through the Web site.
Then if she wants to contact you, she can give you her personal
information.

M: She'll probably think I'm crazy for contacting her almost 25 years later.

D: I'm sure she'll be happy to receive communication from a good friend

_____ hasn't seen in years. When I
(10)

graduate from high school, I'm never going to lose contact with

the friends _____ made. We'll always stay in touch.
(11)

M: That's what you think. But as time passes and your lives become
more complicated, you may lose touch.

D: But today we have e-mail.

M: Well, e-mail is a help. Even so, the direction you _____
(12)

in life is different from the direction your friends choose.

EXERCISE **8** Fill in the blanks with appropriate words to complete the
conversation. Answers may vary.

A: I'm lonely. I have a lot of friends in my native country, but I don't

have enough friends here. The friends I have there send me e-mail
(example)

all the time, but that's not enough. I need to make new friends here.

B: Haven't you met any people here?

A: Of course. But the people _____ here don't have my
(1)

interests.

B: What are you interested in?

A: I like reading, meditating, going for quiet walks. Americans seem to like parties, TV, sports, movies, going to restaurants.

B: You're never going to meet people with the interests _____ . Your interests don't include other people. You
        (2)

should find some interests _____ other people, like
                                  (3)

tennis or dancing, to mention only a few.

A: The activities _____ cost money, and I don't have a lot
                        (4)

of money.

B: There are many parks in this city _____ free tennis courts.
                                            (5)

If you like to dance, I know of a park district near here

_____ free dance classes. In fact, there are a lot of
        (6)

things _____ or very low cost in this city. I can give you a
            (7)

list of free activities, if you want.

A: Thanks. I'd love to have the list. Thanks for all the suggestions ____

_____ .
        (8)

B: I'd be happy to give you more, but I don't have time now. Tomorrow I'll bring you a list of activities from the parks in this city. I'm sure you'll find something _____ on that list.
                                        (9)

A: Thanks.

EXERCISE 9 We often give a definition with an adjective clause. Work with a partner to give a definition of the following words by using an adjective clause.

EXAMPLES twins
Twins are brothers or sisters who are born at the same time.
an answering machine
An answering machine is a device that takes phone messages.

1. a babysitter   5. a fax machine
2. an immigrant   6. a dictionary
3. an adjective   7. a mouse
4. a verb   8. a coupon

mouse

11.4 | *Where* and *When*

Examples	Explanation
Some classes have their reunion in the school **where they first met.** There are Web sites **where you can find lists of high schools and their students.** She attended the University of Washington, **where she met her best friend.**	*Where* means "in that place." *Where* cannot be omitted.
Do you remember the time **(when) you were in high school?** High school was a time **(when)** I had many good friends and few responsibilities. In 1984, **when I graduated from high school**, my best friend's family moved to another state.	*When* means "at that time." *When* can sometimes be omitted.

Punctuation Notes:

1. An adjective clause is sometimes separated from the sentence with a comma. This is true when the person or thing in the main clause is unique.

Compare:

    I visited a Web site **where** I found the names of my classmates. (No Comma)

    I visited Classmates. com, **where** I found the names of my classmates. (Comma: Classmates. com is a unique Web site.)

    I remember the year **when** I graduated from high school. (No Comma)

    In 1984, **when** she graduated from high school, she got married. (Comma: 1984 is a unique year.)

2. *When* without a comma can be omitted.

    I remember the year I graduated from high school.

EXERCISE **10**   This is a conversation between a son (S) and his dad (D). Fill in the blanks with *where* or *when* to complete this conversation.

S: How did you meet mom? Do you remember the place

    <u> where </u> you met?
    (example)

D: We met in high school. I'll never forget the day _____
                                                     (1)

    I met your mother. She was such a pretty girl.

S: Did you go to the same school?

D: Yes. We were in a typing class together. She was sitting at the typewriter next to mine.

S: Dad, what's a typewriter?

D: There was a time _____ we didn't have computers. We had
                                   (2)

    to type our papers on typewriters.

S: Did you start dating right away?

D: No. We were friends. There was a time _____ people were
(3)

friends before they started dating. There was a soda shop near
school _____ we used to meet.
(4)

S: What's a soda shop, Dad?

D: It's a store _____ you could buy milk shakes, sodas, and
(5)

hamburgers. We used to sit there after school drinking one soda
with two straws.

S: That doesn't seem too romantic to me.

D: But it was.

S: So did you get married as soon as you graduated from high school?

D: No. I graduated from high school at a time _____ there was
(6)

a war going on in this country. Mom went to college and I went into the
army. We wrote letters during that time. When I got out of the army, I
started college. So we got married about seven years after we met.

11.5 | Formal vs. Informal

Examples	Explanation
Informal: I lost touch with the friends I used to go to high school **with.** Formal: I lost touch with the friends **with whom** I used to go to high school.	Informally, most native speakers put the preposition at the end of the adjective clause. The relative pronoun is usually omitted.
Informal: I saved the yearbook my friends wrote **in.** Formal: I saved the yearbook **in which** my friends wrote.	In very formal English, the preposition comes before the relative pronoun, and only *whom* and *which* may be used. *That* is not used directly after a preposition.

EXERCISE **11**   Change the sentences to formal English.

EXAMPLE   What is the name of the high school you graduated from?

What is the name of the high school from which you graduated?

1. He found his friend that he served in the military with.

_____

2. I can't find the friend I was looking for.

_____

3. The high school she graduated from was torn down.

_____

4. Do you remember the teacher I was talking about?

_____

5. In high school, the activities I was interested in were baseball and band.

_____

INTERNET MATCHMAKING

Before You
Read

1. **Where do you meet new people?**
2. **Do you know anyone who has tried an online dating service?**

 Read the following article. Pay special attention to adjective clauses beginning with *whose*.

Is it possible to find love on the Internet? About 40 million people a month visit an online dating service in hopes of finding true love. Some of these dating sites let you easily search the pictures and biographical descriptions of people who list themselves there. You can search by age and location. Other Web sites make you fill out lengthy questionnaires so that they can match you with people **whose interests** and **values** are similar to yours. Most of these sites charge a fee for this service.

Meg Olson, a 40-year-old woman from Michigan who wanted to get married, was simply not meeting men. She used an online dating site and met Don Trenton, 42, **whose wife** had recently died. After e-mailing, they started to talk on the phone and realized how many things they had in common. They met, started dating, and a year later, they were married. Don, **whose son** was six years old at the time, wanted to create a stable family for his son.

There are sites for all kinds of interests and connections. Some sites specialize in a specific religion or ethnic group. There are sites for senior citizens. Sadie Kaplan is a 75-year-old widow who wants to meet men **whose age** and **interests** are similar to her own. However, it's harder for women in this age group to meet men because women live longer than men. Because the life expectancy for women is much higher than it is for men (79 for women, 73 for men), many of the women in her age group are widows.

As people live busier and busier lives, they sometimes don't have the time to go out and meet new people. Dating Web sites provide a fast, easy way for people to find romance.

11.6 | *Whose* + Noun

Whose is the possessive form of *who*. It substitutes for *his, her, its, their,* or the possessive form of the noun.

He met a woman.		
	Her	values are similar to his own.
He met a woman	whose	values are similar to his own.

Don wanted to create a family.		
	Don's	son was six years old.
Don,	whose	son was six years old, wanted to create a family.

Language Note:

Use *who* to substitute for a person. Use *whose* for possession or relationship.

Compare:

She married a man **who** has a child.

She married a man **whose interests** are similar to hers.

Punctuation Note:

An adjective clause is sometimes separated from the sentence with a comma. This is true when the person or thing in the main clause is unique.

Compare:

Some dating Web sites match you with someone **whose values** are similar to your own.

Sally met Harry, **whose values** were similar to her own. (Harry is unique.)

EXERCISE 12   This is a conversation between two friends.   Fill in the blanks. Answers may vary.

A: I know you're trying to meet a man. I have a cousin whose
    ___wife___ died last year. He's your age. He's ready to start dating.
    (example)

B: Tell me more about him.

A: He likes sports and the outdoors.

B: You know I don't like sports. I prefer to stay home and read or watch
   movies. I want to meet someone whose _____
                                                        (1)
   are the same as mine.

A: I'm sure you can become interested in football and fishing.

B: I'm not so sure about that. What kind of work does he do?

A: He's a traveling salesman. He's almost never home.

B: I prefer to meet a man whose _____ doesn't take him
                                          (2)
   away from home all the time. What else can you tell me about him?

A: He has a three-year-old son and a five-year-old daughter.

B: I don't want to marry a man whose _____ are small.
    (3)
    My kids are grown up, and I don't want to start raising kids again.

A: It's okay. His mother lives with him now, and she helps take care of the kids.

B: I don't want to date a man whose _____ lives with him.
    (4)

A: But she's a nice woman. She's my aunt.

B: I'm sure she is, but I'm 45 and don't want to live with someone's mother.

A: You know my intentions are good.

B: I have a lot of friends whose _____ are good, but
    (5)
    then I meet the man and find we have nothing in common. I think it's better if I meet a man on my own.

EXERCISE 13   ABOUT YOU Fill in the blanks.

EXAMPLE   I would like to own a car that _has enough room for my large family._

1. My mother is a woman who
2. My city is a place where
3. My childhood was a time when
4. My favorite kind of book is one that
5. A great teacher is a person who
6. I have a friend whose
7. I have a computer that
8. I like to shop at a time when
9. I don't like people who

EXERCISE 14   *Combination Exercise. Part A:* Some women were asked what kind of man they'd like to marry. Fill in the blanks with a response, using the words in parentheses ( ).

EXAMPLE   I'd like to marry a man ___whose values are the same as mine.___
(His values are the same as mine.)

1. I'd like to marry a man _____
   (I can trust him.)

2. I don't want a husband _____
   (He doesn't put his family first.)

3. I want to marry a man _____
   (He makes a good living.)

4. I'd like to marry a man _____
   (His mother lives far away.)

5. I'd like to marry a man _____
   (He's older than I am.)

6. I'd like to marry a man _____
   (He wants to have children.)

7. (Women: Add your own sentence telling what kind of man you'd like to marry, or what kind of man you married.)

_____

Part B: Some men were asked what kind of woman they'd like to marry. Fill in the blanks with a response, using the words in parentheses ( ).

EXAMPLE   I'd like to marry a woman who knows how to cook.
(She knows how to cook.)

1. I'd like to marry a woman _____
   (She has a sense of humor.)

2. I'd like to marry a woman _____
   (I can admire her wisdom.)

3. I'd like to marry a woman _____
   (Her manners are good.)

4. I'd like to marry a woman _____
   (Her family is supportive.)

5. I'd like to marry a woman _____
   (I have known her for a long time.)

6. I'd like to marry a woman _____
   (She wants to have a lot of kids.)

7. (Men: Add your own sentence telling what kind of woman you'd like to marry, or what kind of woman you married.)

_____

EXERCISE 15   *Combination Exercise.* Fill in the blanks with appropriate words to complete the conversation. Answers may vary.

A: I'm getting married in two months.

B: Congratulations. Are you marrying the woman ____you met____
                                                    (example)

   at Mark's party last year?

A: Oh, no. I broke up with that woman a long time ago. I'm going to marry a woman _____ online about ten months ago
(1)

B: What's your fiancée's name? Do I know her?

A: Sarah Liston.

B: I know someone whose _____ is Liston.
(2)

I wonder if they're from the same family.

A: I doubt it. Sarah comes from Canada.

B: Where are you going to live after you get married? Here or in Canada?

A: We're going to live here. Sarah's just finishing college and doesn't have a job yet. This is the place _____
(3)

I have a good job, so we decided to live here.

B: Where are you going to get married?

A: At my parents' friend's house. They have a very big house and garden. The wedding's going to be in the garden.

B: My wife and I made plans to get married outside too, but we had to change our plans because it rained that day.

A: That's OK. The woman _____ is
(4)

more important than the place _____
(5)

we get married. And the life _____ together
(6)

is more important than the wedding day.

B: You're right about that!

EXERCISE 16   *Combination Exercise.* Use the words in parentheses ( ) to form an adjective clause. Then read the sentences and tell if you agree or disagree. Give your reasons.

EXAMPLE   A good friend is a person _____ I can trust. _____
(I can trust her.)

1. A good friend is a person _____ almost every day.
(I see him.)

2. A good friend is a person _____ .
(She would lend me money.)

3. A good friend is a person _____ .
(He knows everything about me.)

4. A person _____ cannot be my friend.
(He has different political opinions.)

5. A person _____ cannot
   (She doesn't speak my native language.)

   be my good friend.

6. A person _____ cannot
   (His religious beliefs are different from mine.)

   be my good friend.

7. A person _____ cannot be a good friend.
   (She lives far away.)

8. I would discuss the problems _____ with a
   (I have problems.)

   good friend.

9. This school is a place _____
   (I can make many new friends easily at this school.)

10. Childhood is the only time in one's life _____

    _____

    (It is easy to make friends at this time.)

SUMMARY OF LESSON 11

Adjective Clauses

1. Pronoun as Subject

   She likes men **who have self-confidence.**

   The man **that arrived late** took a seat in the back.

2. Pronoun as Object

   I'd like to meet the man **(who/m) (that) she married.**

   The book **(which) (that) I'm reading** is very exciting.

3. Pronoun as Object of Preposition

FORMAL:	The person **about whom** I'm talking is my cousin.
INFORMAL:	The person **(who)** I'm talking **about** is my cousin.
FORMAL:	The club **of which** I am a member meets at the community center.
INFORMAL:	The club **(that)** I am a member **of** meets at the community center.

4. *Whose* + Noun

   I have a friend **whose brother lives in Japan.**

   The students **whose last names begin with A or B** can register on Friday afternoon.

5. *Where*

   He moved to New Jersey, **where** he found a job.

   The apartment building **where** he lives has a lot of immigrant families.

6. *When*

　　She came to the U.S. at a time **when** she was young enough to learn English easily.

　　She came to the U.S. in 1995, **when** there was a war going on in her country.

EDITING ADVICE

1. Use *who*, *that*, or *which* to introduce an adjective clause. Don't use *what*.

　　　　　　　　who
　　I know a woman ~~what~~ has ten cats.

2. If the relative pronoun is the subject, don't omit it.

　　　　　　　who
　　I know a man ∧ has been married four times.

3. Use *whose* to substitute for a possessive form.

　　　　　　　　　　whose
　　I live next door to a couple ~~their~~ children make a lot of noise.

4. If the relative pronoun is used as the object, don't put an object after the verb of the adjective clause.

　　I had to pay for the library book that I lost ~~it~~.

5. Don't use *which* for people.

　　　　　　who
　　The man ~~which~~ bought my car paid me by check.

6. Use subject-verb agreement in all clauses.

　　　　　　　　　　　s
　　I have a friend who live ∧ in Madrid.

　　People who talks too much bother me.

7. Don't use an adjective clause when a simple adjective is enough.

　　　I don't like long movies.
　　~~I don't like movies that are long.~~

8. An adjective clause is a dependent clause. It is never a sentence.

　　　　　　　　　　　　　who
　　I sold my car to a man. ~~Who~~ lives on the next block.

9. Put a noun before an adjective clause.

　　　A student w
　　~~Who~~ needs help should ask the teacher.

10. Put the adjective clause immediately after the noun it describes.

The car is beautiful (that you bought).

11. Use *where*, not *that*, to mean "in a place."

where
The store ~~that~~ I buy my textbooks is having a sale this week.

12. Use *whom* and *which*, not *that*, if the preposition precedes the relative pronoun.

which
She would never want to go back to the country from ~~that~~ she came.

13. Use correct word order in an adjective clause (subject before verb).

my father caught
The fish that ~~caught my father~~ was very big.

14. Don't confuse *whose* (possessive form) and *who's (who is)*.

who's
A woman ~~whose~~ in my math class is helping me study for the test.

LESSON 11 TEST/REVIEW

PART **1** Find the mistakes with the underlined words, and correct them. Not every sentence has a mistake. If the sentence is correct, write *C*.

who's
EXAMPLES   Do you know the man ~~whose~~ standing in the back of the theater?

Could you please return the book I lent you last week?   C

1. The wallet which found my friend has no identification.
2. The coat is too small that I bought last week.
3. I don't know the people who lives next door to me.
4. I have to return the books that I borrowed from the library.
5. I don't like neighbors what make a lot of noise.
6. I don't like the earrings that I bought them.
7. I have a friend lives in Houston.
8. Who speaks English well doesn't have to take this course.
9. I can't understand a word you are saying.
10. I prefer to have an English teacher which speaks my language.
11. Everyone whose last name begins with A should stand up.

12. The store <u>that</u> I buy my groceries is open 24 hours a day.

13. I don't understand a thing <u>you are talking about</u>.

14. The woman <u>with whom he came to the party</u> was not his wife.

15. I don't know <u>anyone. Who</u> has a record player my more.

16. We rented an apartment <u>that doesn't have</u> a refrigerator.

17. A couple <u>who's</u> children are small has a lot of responsibilities.

18. I have a friend <u>her</u> brother just graduated from medical school.

PART **2**   Fill in the blanks to complete the adjective clause.   Answers may vary.

EXAMPLE   A: You lost a glove. Is this yours?

B: No. The glove ___that I lost___ is brown.

1. A: My neighbor's children make a lot of noise.

   B: That's too bad. I don't like to have neighbors _____

   _____

2. A: I have a new cat. Do you want to see him?

   B: What happened to the other cat _____

   A: She died last month.

3. A: Do you speak French?

   B: Yes, I do. Why?

   A: The teacher is looking for a student _____
   to help her translate a letter.

4. A: Did you meet your boyfriend on an Internet dating site?

   B: No. I didn't like any of the men _____
   on the Internet.

5. A: Does your last name begin with *A*?

   B: Yes, it does. Why?

   A: Registration is by alphabetical order. Students _____

   _____ can register after two o'clock today.

6. A: Did you go to your last high school reunion?

   B: No. I was out of town on the day _____.

   A: Do you usually go to your reunions?

   B: Yes. I love to keep in touch with the people _____.

7. A: Are you planning to marry Charles?

   B: No. He lives with his mother. I want to marry a man _____

   _____ lives far away.

EXPANSION ACTIVITIES

Classroom Activities

1. Tell if you agree or disagree with the statements below. Discuss your answers.

	I agree.	I disagree.
a. People who have different religions can have a good marriage.		
b. People who come from different countries or have different languages can have a good marriage.		
c. Women who marry younger men can be happy.		
d. It's possible to fall in love with someone you've just met.		
e. Young people who want to get married should get the approval of their parents.		
f. A man shouldn't marry a divorced woman who has children.		
g. Couples who have children shouldn't get divorced.		
h. Older women whose husbands have died should try to get married again.		
i. A man should always marry a woman who is shorter than he is.		
j. Couples who live with a mother-in-law usually have problems.		
k. A woman shouldn't marry a man who has a lower level of education.		

2. Write a short definition or description of an object or a person. Read your definition to a small group. The others will try to guess what it is. Continue to add to your definition until someone guesses it.

EXAMPLE   It's an animal that lives in the water.
           Is it a fish?
           No, it isn't. It's an animal that needs to come up for air.
           Is it a dolphin?
           Yes, it is.

3. Write a word from your native language that has no English translation. It might be the name of a food or a traditional costume. Define the word. Read your definition to a small group or to a partner.

EXAMPLE   A *sari* is a typical Indian dress for women. It is made of a cloth that a woman wraps around her. She wraps one end around her waist. She puts the other end over her shoulder.

4. Bring to class something typical from your country. Demonstrate how to use it.

EXAMPLE   a samovar
           This is a pot that we use in Russia to make tea.

5. Dictionary game. Form a small group. One student in the group will look for a hard word in the dictionary. (Choose a noun. Find a word that you think no one will know.) Other students will write definitions of the word. Students can think of funny definitions or serious ones. The student with the dictionary will write the real definition. Students put all the definitions in a box. The student with the dictionary will read the definitions. The others have to guess which is the real definition.

EXAMPLE   parapet

Sample definition: A parapet is a small pet that has wings, like a parakeet.

Real definition: A parapet is a low wall that runs along the edge of a roof or balcony.

(The teacher can provide a list of words and definitions beforehand, writing them on small pieces of paper. A student can choose one of the papers that the teacher has prepared.)

1. Do you think the Internet is a good way to meet a romantic partner? Why or why not?

2. How do people in your native culture find a spouse?

3. Talk about the kind of person who makes a good husband, wife, father, mother, or friend.

4. If you are married, tell where or how you met your spouse.

5. Are you surprised that there are Internet sites for seniors who are single and looking for a partner?

6. In your native culture, do people usually keep in touch with the friends they made in school?

7. Are there class reunions in your native country?

1. Write a short composition describing your best friend from your school days.

2. Write a short composition describing the difference between dating customs in the U.S. and in your native culture.

1. Visit an online dating service. Bring in a profile of a person that you think is interesting.

2. Visit an online dating service for senior citizens. Bring in a profile of a person that you think is interesting.

3. Visit a Web site that lists classmates. If you graduated from high school in the U.S., see if your high school is listed.

4. Visit a Web site that plans reunions. Find out some of the steps that are necessary in planning a reunion.

Additional Activities at http://elt.thomson. com/gic

LESSON
12

GRAMMAR

Superlatives

Comparatives

CONTEXT: Sports and Athletes

Michael Jordan
Americans' Attitude Toward Soccer
An Amazing Athlete
Football and Soccer

12.1 | Superlatives and Comparatives—An Overview

Examples	Explanation
Baseball and basketball are **the most popular** sports in the U.S.	We use the superlative form to point out the number one item or items in a group of three or more.
Jack is **the tallest player** on the basketball team.	
Baseball is **more popular than** soccer in the U.S. Basketball players are **taller than** baseball players.	We use the comparative form to compare two items or groups of items.
He is **as tall as** a basketball player. Soccer is not **as popular** as baseball. Soccer players are not **the same height as** basketball players.	We can show equality or inequality.

MICHAEL JORDAN

Before You **Read**

1. Do you like sports? Which are your favorites?

2. Who are your favorite athletes?

Read the following article. Pay special attention to superlative forms.

Michael Jordan is probably **the best known** basketball player in the world. His career started in the early 1980s, when he played college basketball with the University of North Carolina. Although he was not **the tallest** or **the strongest** player, he won the attention of his coach for being an excellent athlete. Probably his **most important** achievement[1] at that time was scoring the winning basket in the 1983—1984 college championship game.

Jordan left college early to join the Chicago Bulls, a professional basketball team. He led the Bulls to **the best** record in professional basketball history. He was voted **the most valuable** player five times. Jordan holds several records: **the highest** scoring average (31. 7 points per game) and **the most** points in a playoff game (63). Many people think that Michael Jordan was **the most spectacular** basketball player of all time. A statue of Jordan in Chicago has these words, "**The best** there ever was. **The best** there ever will be." Jordan retired from the Chicago Bulls in 1999 at the age of 35.

Jordan came out of retirement in 2001 to play a few more seasons with another team, the Washington Wizards. When he announced his comeback, he said he would donate his $ 1 million salary the first year to the families of the victims of the September 11, 2001 terrorist attacks. He retired from basketball for good in 2003, at the age of 40. L.A. Laker superstar Magic Johnson said it well when he said, "There's Michael, then there's all the rest of us."

Jordan's popularity with his fans brought him to the attention of advertisers. Jordan is paid a lot of money to have his name appear on sports products and to appear in TV commercials. *Forbes Magazine* in 2004 ranked him the fourth **highest** paid athlete and the seventh **highest** paid celebrity.

In his retirement, Jordan works with charities. He created The Jordan Institute for Families, an organization that tries to help solve the problems facing poor families. He is hoping to help families accomplish their dreams.

[1]An *achievement* is something you attain through practice or hard work.

12.2 | The Superlative Form

We use the superlative form to point out the number one item of a group of three or more. The superlative has two forms, depending on the number of syllables in the adjective or adverb.

Examples	Explanation
Jordan was not **the tallest** player on his team. Jordan was not **the strongest** player on his team.	Use: *the* + [short adjective/adverb] + *-est* We often put a prepositional phrase after a superlative phrase: *in the world, on his team, in the U.S., of all time.*
Jordan was probably **the most spectacular** player of all time. He was **the most valuable** player on this team.	Use: *the most* + [long adjective/adverb]
Jordan is **one of the richest athletes** in the world. Jordan was **one of the oldest players** on his team.	We often say "one of the" before a superlative form. The noun that follows is plural.
Jordan is one of the best athletes **who has ever lived**. His last game with the Bulls was one of the most exciting games **I have ever seen**.	An adjective clause with *ever* and the present perfect tense often completes a superlative statement.
Who is **the best** athlete in the world?	Some superlatives are irregular. See 12.3 for more information.

Language Note:

Use **the** before a superlative form. Omit **the** if there is a possessive form before the superlative form.

Jordan was **the team's most valuable** player. (*Not:* Jordan was the *team's the most* valuable player.)

My oldest brother loves basketball. (*Not:* My the oldest brother loves basketball.)

EXERCISE **1**   Tell if the statement is true *(T)* or false *(F)*. Underline the superlative forms. Not every sentence has a superlative form.

EXAMPLE   Michael Jordan is one of <u>the best</u> basketball players in the world.   T

1. Magic Johnson said, "Jordan is the best there ever was, the best there ever will be."
2. Jordan was voted the most valuable player more than one time.
3. Jordan is one of the richest athletes in the world.
4. Jordan scored the most points in a playoff game.
5. Jordan retired for good at the age of 35.
6. One year, Jordan donated his $1 million dollar salary to the families of the victims of September 11.

12.3 | Comparative and Superlative Forms of Adjectives and Adverbs

Explanation	Simple	Comparative	Superlative
One-syllable adjectives and adverbs*	tall	taller	the tallest
	fast	faster	the fastest
Two-syllable adjectives that end in -y	easy	easier	the easiest
	happy	happier	the happiest
Other two-syllable adjective	frequent	more frequent	the most frequent
	active	more active	the most active
Some two-syllable adjectives have two forms.	simple	simpler	the simplest
		more simple	the most simple
Other two-syllable adjectives that have two forms are *handsome, quiet, gentle, narrow, clever, friendly, angry, polite, stupid.*	common	commoner	the commonest
		more common	the most common
Adjectives with three or more syllables	important	more important	the most important
	difficult	more difficult	the most difficult
-ly adverbs	quickly	more quickly	the most quickly
	brightly	more brightly	the most brightly
Irregular adjectives and adverbs	good / well	better	the best
	bad / badly	worse	the worst
	far	farther	the farthest
	little	less	the least
	a lot	more	the most

Spelling Rules for short Adjectives and Adverbs			
Rule	Simple	Comparative	Superlative
Add -er and -est to short adjectives and adverbs.	tall	taller	tallest
	fast	faster	fastest
For adjectives that end in *y*, change *y to i* and add -er and -est.	easy	easier	easiest
	happy	happier	happiest
For adjectives that end in e, add -r and -st.	nice	nicer	nicest
	late	later	latest
For words ending in consonant-vowel-consonant, double the final consonant, then add -er and -est.**	big	bigger	biggest
	sad	sadder	saddest

Language Notes:
*Exceptions: bored   more bored   the most bored
                    tired   more tired   the most tired
**Exceptions: Do not double final *w:* new—newer—newest

EXERCISE **2**   Give the comparative and superlative forms of each word.

EXAMPLES   fat _____ fatter _____   _____ the fattest _____

important _____ more important _____   _____ the most important _____

1. interesting _____ _____
2. young _____ _____
3. beautiful _____ _____
4. good _____ _____
5. common _____ _____
6. thin _____ _____
7. carefully _____ _____
8. pretty _____ _____
9. bad _____ _____
10. famous _____ _____
11. lucky _____ _____
12. simple _____ _____
13. high _____ _____
14. delicious _____ _____
15. far _____ _____
16. foolishly _____ _____

EXERCISE **3**   Many people have said that Jordan is or was the superlative in these categories. Write the superlative form in each blank.

EXAMPLE   He was _____ the most elegant _____ athlete.
                              (elegant)

1. He was _____ athlete.
                        (popular)

2. He was _____ athlete.
                        (great)

3. He was _____ athlete.
                        (powerful)

4. He was _____ athlete.
                        (graceful)

5. He is _____ -known American basketball
                    (good)

player in the world.

6. He was _____ player.
   (valuable)

7. He is one of _____ people in the world.
   (rich)

8. He is one of _____ -dressed people in the world.
   (good)

EXERCISE **4**   Write the superlative form of the word in parentheses ( ).

1. Michael Schumacher is one of _____
   (fast)

   race car drivers in the world.

2. Training for the Olympics is one of _____
   (difficult)

   things for an athlete.

3. Soccer is _____ sport in the world.
   (popular)

4. Sumo wrestlers are _____ athletes.
   (fat)

5. Michael Jordan was _____ player on the
   (valuable)

   Chicago Bulls.

6. Swimming and gymnastics are _____ events
   (watched)

   during the Summer Olympics.

7. Yao Ming is one of _____
   (tall)

   basketball players in the world.

8. _____ name for
   (common)

   soccer in the world is "football."

9. Running a marathon was one of _____
   (hard)

   things I've ever done.

10. In your opinion, what is _____

    _____ sport?
    (interesting)

EXERCISE **5**   ABOUT YOU Write a superlative sentence giving your opinion about each of the following items. You may find a partner and compare your answers to your partner's answers.

EXAMPLES   **big problem in the world today**
I think the biggest problem in the world today is hunger.

big problem in the U.S. today
I think crime is one of the biggest problems in the U.S. today.

1. good way to make friends

   _____

2. quick way to learn a language

   _____

3. good thing about life in the U.S.

   _____

4. bad thing about life in the U.S.

   _____

5. terrible tragedy in the world

   _____

6. big problem in *(choose a country)*

   _____

EXERCISE **6**   ABOUT YOU Write superlative sentences about your experience with the words given.   Use the present perfect form after the superlative.

EXAMPLE   **big / city / visit**
London is the biggest city I have ever visited.

1. **tall / building / visit**

_____

2. **beautiful / actress / see**

_____

3. **difficult / subject / study**

_____

4. **far / distant / travel**

_____

5. **bad / food / eat**

_____

6. **good / vacation / have**

_____

7. **good / athlete / see**

_____

8. **hard / job / have**

_____

9. **interesting / sporting event / see**

_____

EXERCISE **7**   ABOUT YOU Fill in the blanks.

EXAMPLE   *Swimming across a lake alone at night*
was one of the most dangerous things I've ever done.

1. _____
   is one of the most foolish things I've ever done.

2. _____
   is one of the hardest decisions I've ever made.

3. _____
   is one of the most dangerous things I've ever done.

12.4 | Superlatives and Word Order

Examples	Explanation
Superlative Noun Adjective Phrase Who is **the best basketball player?** Superlative Adjective Noun Who is **the most popular player?**	A superlative adjective comes **before** a noun or noun phrase.
Football is **the most popular sport** in the U.S. OR **The most popular sport** in the U.S. is football.	When the verb *be* connects a noun to a superlative adjective + noun, there are two possible word orders.
Verb Superlative Adverb Interest in soccer **is growing the most quickly** in the U.S. Verb Phrase Superlative Adverb Michael Jordan **shot baskets the most gracefully.**	We put superlative adverbs **after** the verb (phrase).
Verb Superlative Michael jordan **played the best** with the Bulls. Verb Phrase Superlative Fans **loved Michael Jordan the most.**	We put *the most, the least, the best,* and *the worst* **after** a verb (phrase).
Superlative Noun Who scored **the most points?** Superlative Noun The Bulls had **the best record.**	We put *the most, the least, the best,* and *the worst* **before** a noun.

EXERCISE **8** ABOUT YOU Name the person who is the superlative in your family in each of the following categories.

EXAMPLE   works hard

My mother works the hardest in my family.

1. drives well
2. lives far from me
3. speaks English confidently
4. spends a lot of money
5. is well dressed
6. watches a lot of TV
7. worries a lot
8. lives well
9. works hard
10. is athletic
11. is a big sports fan
12. is learning English quickly

AMERICANS' ATTITUDE TOWARD SOCCER

Before You Read

1. Are you interested in soccer?

2. What's your favorite team?

 Read the following article. Pay special attention to comparisons.

Soccer is by far the most popular sport in the world. Almost every country has a professional league. In many countries, top international soccer players are **as** well-known **as** rock stars or actors. However, in 1994 when the World Cup soccer competition was held in the U.S., there was not a lot of interest in soccer among Americans. Many people said that soccer was boring.

Recently, Americans' attitude toward soccer has been changing. In 1999, when the Women's World Cup was played in the U.S., there was **more** interest than ever before. Little by little, soccer is becoming **more popular** in the U.S. The number of children playing soccer is growing. In fact, soccer is growing **faster** than any other sport. For elementary school children, soccer is now the number two sport after basketball. **More** kids play soccer than baseball. Many coaches believe that soccer is **easier** to play than baseball or basketball, and that there aren't **as many** injuries **as** with sports such as hockey or football.

Interest in professional soccer in the U.S. is still much **lower** than in other countries. The number of Americans who watch professional basketball, football, or hockey is still much **higher** than the number who watch Major League Soccer. However, **the more** parents show interest in their children's soccer teams, **the more** they will become interested in professional soccer.

12.5 | Comparatives

We use the comparative form to compare two items. The comparative has two forms, depending on the number of syllables in the adjective or adverb.

Examples	Explanation
Soccer players are **shorter than** basketball players. Interest in baseball is **higher than** interest in soccer in the U.S.	Use: short adjective/short adverb + -er + than

Basketball is **more popular than** soccer in the U.S.
Interest in soccer is growing **more quickly than** interest in hockey.
	Use:
	  more + longer adjective + than
	  more + -ly adverb + than

My brother plays soccer **better than** I do.	Some comparative forms are irregular. See 12.3 for more information.
Basketball is popular in the U.S., but football is **more popular**. Michael Jordan is tall, but other basketball players are **taller**.	Omit than if the second item of comparison is not included.
Interest in soccer is *much* **lower** in the U.S. than in other countries. I like soccer *a little* **better** than I like baseball.	*Much* or *a little* can come before a comparative form.
You are taller than **I am.** (FORMAL) You are taller than **me.** (INFORMAL) I can play soccer better than **he can.** (FORMAL) I can play soccer better than **him.** (INFORMAL)	When a pronoun follows *than*, the correct form is the subject pronoun (*he, she, I,* etc.). Usually an auxiliary verb follows (*is, do, did, can,* etc.). Informally, many Americans use the object pronoun (*him, her, me,* etc.) after *than*. An auxiliary verb does not follow.
The more they practice, **the better** they play. **The older** you are, **the harder** it is to learn a new sport.	We can use two comparisons in one sentence to show cause and result.

EXERCISE 9   Circle the correct word to complete each statement.

EXAMPLE   In the U.S., soccer is *more / (less)* popular than basketball.

1. Football players have *more / fewer* injuries than soccer players.

2. In the U.S., soccer is growing *faster / slower* than any other sport.

3. In 1999, there was *more / less* interest in soccer than in 1994.

4. Professional soccer is *more / less* popular in the U.S. than in other countries.

5. In the U.S., soccer players are *more / less* famous than movie stars.

EXERCISE **10**   Fill in the blanks with the comparative form of the word in parentheses ( ).

EXAMPLE   In the U.S., basketball is _____more popular than_____ soccer.
                                        (popular)

1. Tall people are often _____ basketball players
                          (good)
   _____ short people.

2. Do you think volleyball is _____ tennis?
                               (fun)

3. Which do you think is _____, skiing or surfing?
                          (difficult)

4. A soccer ball is _____ a tennis ball.
                     (large)

5. Children learn sports _____ adults.
                          (easily)

6. People who exercise a lot are in _____ shape
                                     (good)
   _____ people who don't.

7. Do you think soccer is _____ football?
                           (interesting)

8. Do you think soccer is _____ than baseball?
                           (exciting)

EXERCISE **11**   ABOUT YOU Compare the people of your native country (or a place you know well) to Americans (in general). Give your own opinion.

EXAMPLE   tall

Americans are taller than Koreans.

1. polite	4. tall	7. wealthy
2. friendly	5. thin	8. educated
3. formal	6. serious	9. happy

ABOUT YOU Compare the U.S. and your native country (or a place you know well). Explain your response.

EXAMPLES cars
Cars are cheaper in the U.S. Most people in my native country can't afford a car.
education
Education is better in my native country. Everyone must finish high school.

1. rent	4. education	7. gasoline
2. housing	5. medical care	8. the government
3. cars	6. food	9. clothes (or fashions)

12.6 | Comparatives and Word Order

Examples	Explanation
Comparative Be Adjective Basketball **is more popular** than soccer in the U.S. Linking Comparative Verb Adjective Football **looks more dangerous** than soccer.	put the comparative adjective after the verb be or other linking verbs: *seem, feel, look, sound, etc.*
Verb Phrase Comparative Adverb Jordan **played basketball more gracefully** than any other player. comparative verb Adverb Soccer **is growing faster** than any other sport.	put the comparative adverb **after** the verb (phrase).
Comparative Noun There is **less interest** in hockey than there is in basketball. Comparative Noun Soccer players have **fewer injuries** than football players.	We can put *more, less, fewer, better*, and *worse* **before** a noun.
Verb Phrase Comparative My sister **likes soccer more** than I do. Verb Phrase Comparative **I play soccer worse** than my sister does.	You can put *more, less, better*, and *worse* **after** a verb (phrase).

EXERCISE **13** Find the mistakes with word order and correct them. Not every sentence has a mistake. If the sentence is correct, write *C*.

EXAMPLES A football team has players (more) than a baseball team.
A golf ball is smaller than a tennis ball. C

1. A basketball player is taller than a gymnast.
2. A baseball game has action less than a soccer game.

3. Football players use padding more than soccer players.
4. Michael Jordan more beautifully played basketball than other players.
5. I more like baseball than basketball.
6. Team A won more games than Team B.
7. Team A better played than Team B.

EXERCISE 14   ABOUT YOU Use a comparative adverb to compare the people of your native country (or a place you know well) to Americans (in general). Give your own opinion.

EXAMPLE   drive well
Mexicans drive better than Americans.

1. dress stylishly   4. live long   7. have freedom
2. work hard   5. worry a little   8. have a good life
3. spend a lot   6. live comfortably   9. exercise a lot

EXERCISE 15   ABOUT YOU Compare this school to another school you attended. Use *better, worse, more, less, or fewer* before the noun.

EXAMPLE   classroom/space
This classroom has more space than a classroom in my native country.

1. class / students   4. library / books
2. school / courses   5. school / facilities[2]
3. teachers / experience   6. school / teachers

EXERCISE 16   *Combination Exercise.* Fill in the blanks with the comparative or superlative form of the word in parentheses ( ). Include *than* or *the* when necessary.

EXAMPLES   In the U.S., baseball is ___more popular than___ soccer.
                                        (popular)

Baseball is one of ___the most popular___ sports in the U.S.
                                     (popular)

1. A tennis ball is _____ a baseball.
                       (soft)

2. An athlete who wins the gold medal is _____ athlete
                                       (good)
in his or her sport.

3. Who is _____ player on the Chicago Bulls today?
                 (tall)

4. I am _____ in baseball _____ in basketball.
      (interested)

5. In my opinion, soccer is _____ sport.
                         (exciting)

6. Weightlifters are _____ than golfers.
                  (muscular)

[2]*Facilities* are things we use, such as a swimming pool, cafeteria, library, exercise room, or student union.

7. Golf is a _____ sport _____ soccer.
   (slow)

8. A basketball team has _____ players _____
   (few)

   a baseball team.

9. Even though January is _____ month of the year,
   (cold)

   football players play during this month.

10. My friend and I both jog. I run _____ than my friend.
    (far)

11. Who's a _____ soccer player—you or your brother?
    (good)

AN AMAZING ATHLETE

Before You
Read

1. Can people with disabilities do well in sports?

2. Why do people want to climb the tallest mountain in the world?

 Read the following article. Pay special attention to comparisons.

Did You Know?

The oldest person to climb Mount Everest was 70 years old.

Erik Weihenmayer is **as tough as** any mountain climber. In 2001 he made his way to the top of the highest mountain in the world—Mount Everest—at the age of 33. But Erik is **different from** other mountain climbers in one important way—he is completely blind. He is the first sightless person to reach the top of the tallest mountain.

Erik was an athletic child who lost his vision in his early teens. At first he refused to use a cane or learn Braille, insisting he could do **as well as** any teenager. But he finally came to accept his disability and to excel within it. He couldn't play **the same** sports **as** he used to. He would never be able to play basketball or catch a football again. But then he discovered wrestling, a

Know?

Almost 90 percent of Everest climbers fail to reach the top. At least 180 have died while trying.

sport where sight was not **as** important **as** feel and touch. Then, at 16, he discovered rock climbing, which **was like** wrestling in some ways; a wrestler and a rock climber get information through touch. Rock climbing led to mountain climbing, the greatest challenge of his life.

Teammates climbing with Erik say that he isn't **different from** a sighted mountaineer. He has **as much** training **as** the others. He is **as** strong **as** the rest. The major difference is he is not **as** thin **as** most climbers. But his strong upper body, flexibility, mental toughness, and ability to tolerate physical pain make him a perfect climber. The only accommodation for Erik's blindness is to place bells on the jackets of his teammates so that he can follow them easily.

Climbing Mount Everest was a challenge for every climber on Erik's team. The reaction to the mountain air for Erik was **the same as** it was for his teammates: lack of oxygen causes the heart to beat slower than usual and the brain does not function **as clearly as** normal. In some ways, Erik had an advantage over his teammates: as they got near the top, the vision of all climbers was restricted. So at a certain altitude, all his teammates **were like** Erik—nearly blind.

To climb Mount Everest is an achievement for any athlete. Erik Weihenmayer showed that his disability wasn't **as important as** his ability.

12.7 | *As... As*

Examples	Explanation
Erik is **as strong as** his teammates. At high altitudes, the brain doesn't function **as clearly** as normal. Erik can climb mountains **as well as** sighted climbers.	We can show that two things are equal or unequal in some way by using: *as* + adjective / adverb + *as*.
Erik is not **as thin as** most climbers. Skiing is not **as difficult as** mountain climbing.	When we make a comparison of unequal items, we put the lesser item first.
Baseball is popular in the U.S. Soccer is not **as popular.**	Omit the second *as* if the second item of comparison is omitted.

Usage Notes:

1. A very common expression is *as soon as possible*. Some people say *A.S.A.P* for short.

    I'd like to see you *as soon as possible*.

    I'd like to see you *A.S.A.P.*

2. These are some common expressions using *as... as.*

    as poor as a church mouse   as sick as a dog

    as old as the hills   as proud as a peacock

    as quiet as a mouse   as gentle as a lamb

    as stubborn as a mule   as happy as a lark

mule

peacock

EXERCISE **17** Write true (*T*) or false (*F*).

EXAMPLE In wrestling, the sense of sight is as important as the sense of touch. F

1. Rock climbing is not as dangerous as mountain climbing.
2. At high altitudes, you can't think as clearly as you can at lower altitudes.
3. Erik was not as strong as his teammates.
4. When Erik became blind, he wanted to do as well as any other teenager.
5. Erik could not go as far as his teammates.
6. Erik was as prepared for the climb as his teammates.

EXERCISE **18** ABOUT YOU Compare yourself to another person. (Or compare two people you know.) Use the following adjectives and *as...as*. You may add a comparative statement if there is inequality.

EXAMPLES thin

I'm not as thin as my sister. (She's thinner than I am.)

old

My mother is not as old as my father. (My father is older than my mother.)

1. old	4. patient	7. religious	10. talkative
2. educated	5. lazy	8. friendly	11. athletic
3. intelligent	6. tall	9. strong	12. interested in sports

EXERCISE **19** ABOUT YOU Use the underlined word to compare yourself to the teacher.

EXAMPLE speak Spanish <u>well</u>

The teacher doesn't speak Spanish as well as I do. (I speak Spanish better.)

1. arrive at class <u>promptly</u>
2. work <u>hard</u> in class
3. understand American customs <u>well</u>
4. speak <u>quietly</u>
5. speak English <u>fluently</u>
6. understand a foreigner's problems <u>well</u>
7. write <u>neatly</u>
8. speak <u>fast</u>

12.8 | *As Many/Much... As*

Examples	Explanation
Soccer players don't have **as many injuries as** football players. Erik had **as much training as** his teammates.	We can show that two things are equal or not equal in quantity by using *as many* + count noun + *as* or *as much* + noncount noun + *as*.
I don't play soccer **as much as** I used to. She doesn't like sports **as much as** her husband does.	We can use *as much as* after a verb phrase.

EXERCISE 20   ABOUT YOU *Part* A: Fill in the blanks.

EXAMPLE   I drive about _____30_____ miles a week.
                               (number)

1. I'm _____ tall.
          (feet/inches)

2. The highest level of education that I completed is _____

_____.
   (high school, bachelor's degree, master's degree, doctorate)

3. I work _____ hours a week.
            (number)

4. I study _____ hours a day.
             (number)

5. I exercise _____ days a week.
                (number)

6. I'm taking _____ courses now.
                (number)

7. I have _____ siblings.[3]
            (number)

8. I live _____ miles from this school.
            (number)

Part B: Find a partner and compare your answers to your partner's answers. Write statements with the words given and *(not) as...as* or *(not) as much / many as.*

EXAMPLE   **drive** I don't drive as much as Lisa._____

1. **tall** _____

2. **have education** _____

3. **work** _____

4. **study** _____

[3]*Siblings* are a person's brothers and sisters.

5. exercise frequently _____

6. take courses _____

7. have siblings _____

8. live far from school _____

EXERCISE **21** Compare men and women (in general). Give your own opinion. Use *as many as* or *as much as*.

EXAMPLE show emotion

Men don't show as much emotion as women. (Women show more emotion than men.)

1. earn   5. use bad words
2. spend money   6. have responsibilities
3. talk   7. have freedom
4. gossip   8. have free time

EXERCISE **22** ABOUT YOU Compare this school and another school you attended. Use *as many* as.

EXAMPLE classrooms

This school doesn't have as many classrooms as King College. (King College has more classrooms.)

1. teachers   3. floors (or stories)   5. exams
2. classrooms   4. English courses   6. students

EXERCISE **23** Make a comparison between this city and another city you know well using the categories below.

EXAMPLE public transportation The buses are cleaner in Boston than in this city. OR The buses in this city are not as crowded as the buses in Boston.

1. traffic _____

2. people _____

3. gardens and parks _____

4. public transportation _____

5. museums _____

6. universities _____

7. houses _____

8. buildings _____

9. stores or shopping _____

12.9 | *The Same...As*

Examples	Explanation
Pattern A: Erik had **the same ability as** his teammates. A soccer ball isn't **the same shape as** a football.	We can show that two things are equal or not equal in some way by using *the same* + noun + *as*.
Pattern B: Erik and his teammates had **the same ability**. A soccer ball and a football aren't **the same shape**.	Omit *as* in Pattern B.

Language Note:
We can make statements of equality with many nouns, such as *size, shape, color, value, religion,* or *nationality*.

EXERCISE 24   Make statements with *the same...as* using the words given.

EXAMPLES   a golf ball / a tennis ball (size)
A golf ball isn't the same size as a tennis ball.

1. a soccer ball / a volleyball (shape)

_____

2. a soccer player / a basketball player (height)

_____

3. an amateur athlete / a professional athlete (ability)

_____

4. a soccer player / a football player (weight)

_____

5. team A's uniforms / team B's uniforms (color)

_____

EXERCISE 25   ABOUT YOU Talk about two relatives or friends of yours. Compare them using the words given.

EXAMPLE   age
My mother and my father aren't the same age.
OR
My mother isn't the same age as my father.   (My father is older than my mother.)

1. age   3. weight   5. religion
2. height   4. nationality   6. (have) level of education

EXERCISE 26   ABOUT YOU Work with a partner.   Make a **true** affirmative or negative statement with the words given.

EXAMPLES   the same nationality
I'm not the same nationality as Alex. I'm Colombian, and he's Russian.
the same color shoes
Martina's shoes are the same color as my shoes. They're brown.

1. the same hair color
2. the same eye color
3. (speak) the same language

4. (like) the same sports
5. (have) the same level of English
6. the same nationality

12.10 | Equality with Nouns or Adjectives

For equality with nouns, use *the same...as*. For equality with adjectives and adverbs, use *as... as*.

Noun	Adjective	Examples
height	tall, short	A soccer player is not **the same height as** a basketball player. A soccer player is **shorter.**
age	old, young	He's not **the same age as** his wife. His wife is **older.**
weight	fat, thin	Wrestler A is not **the same weight as** wrester B. Wrestler B is **fatter.**
length	long, short	This shelf is not **the same length as** that shelf. This shelf is **shorter.**
price	expensive, cheap	This car is not **the same price as** that car. This car is **cheaper.**
size	big, small	These shoes are not **the same size as** These shoes are **smaller.**

EXERCISE 27 Change the following to use the comparative form. Answers may vary.

EXAMPLE Lesson 11 is not the same length as Lesson 12.

Lesson 11 is _____ shorter _____.

1. I am not the same height as my brother.

My brother is _____.

2. You are not the same age as your husband.

You are _____.

3. I am not the same height as a basketball player.

A basketball player is _____.

4. My left foot isn't the same size as my right foot.

My right foot is _____.

5. My brother is not the same weight as I am.

My brother is _____.

FOOTBALL AND SOCCER

Before You
Read

1. Which do you like better, football or soccer?

2. How are soccer players different from football players?

Read the following article. Pay special attention to similarities and differences.

tackle

It may seem strange that Americans give the name "football" to a game played mostly by throwing and carrying a ball with one's hands. But Americans give the name football to a sport that is very **different from** soccer.

Many of the rules in soccer and American football are the **same**. In both games, there are 11 players on each side, and a team scores its points by getting the ball past the goal of the other team. The playing fields for both teams are also very much **alike**.

When the action begins, the two games look very **different**. In addition to using their feet, soccer players are allowed to hit the ball with their heads. In football, the only person allowed to touch the ball with his feet is a special player known as the kicker. Also, in football, tackling the player who has the ball is not only allowed but encouraged, whereas tackling any player in soccer will get the tackler thrown out of the game.

(continued)

Football players and soccer players don't **dress alike** or even **look alike** in many ways. Since blocking and tackling are a big part of American football, the players are often very large and muscular and wear heavy padding and helmets. Soccer players, on the other hand, are usually thinner and wear shorts and polo shirts. This gives them more freedom of movement to show off the fancy footwork that makes soccer such a popular game around the world.

While both games are very **different**, both have a large number of fans that enjoy the exciting action.

12.11 | Similarity with *Like* and *Alike*

We can show that two things are similar (or not) with *like* and *alike*.

Examples	Explanation
Pattern A: A soccer player **looks like** a rugby player. A soccer player doesn't **dress like** a football player.	Pattern A: Noun 1 + verb + *like* + Noun 2
Pattern B: A soccer player and a rugby player **look alike**. A soccer player and a football player don't **dress alike**.	Pattern B: Noun 1 + Noun 2 + verb + *alike*
Language Note: We often use the sense perception verbs (*look, sound, smell, taste, feel,* and *seem*) with *like* and *alike*. We can also use other verbs with *like: act like, sing like, dress like,* etc.	

EXERCISE **28** Make a statement with the words given.

EXAMPLE taste / Pepsi / Coke

Pepsi tastes like Coke (to me).

OR

Pepsi and Coke taste alike (to me).

1. taste / diet cola / regular cola
2. taste / 2% milk / whole milk
3. look / an American classroom / a classroom in another country
4. sound / Asian music / American music
5. feel / polyester / silk
6. smell / cologne / perfume
7. look / salt / sugar
8. taste / salt / sugar
9. act / American teachers / teachers in other countries
10. dress / American teenagers / teenagers in other countries

EXERCISE **29** Fill in the blanks. In some cases, more than one answer is possible.

EXAMPLE Players on the same team dress _____alike_____.

1. Twins _____ alike.
2. Americans and people from England don't sound _____.
   They have different accents.
3. My daughter is only 15 years old, but she _____ an
   adult. She's very responsible and hard-working.
4. My son is only 16 years old, but he _____ an adult.
   He's tall and has a beard.
5. Teenagers often wear the same clothing as their friends. They like
   to _____.
6. Soccer players don't look _____ football players at
   all.
7. Do you think I'll ever _____ an American, or will I
   always have an accent?

8. Children in private schools usually wear a uniform. They

   _____ alike.

9. My children learned English very quickly. Now they sound

   _____ Americans. They have no accent at all.

10. Dogs don't _____ cats at all. Dogs are very friendly.
    Cats are more distant.

12.12 | *Be Like*

We can show that two things are similar (or not) in internal characteristics with *be like* and be *alike*.

Explanation	Examples
Pattern A: For Erik, mountain climbing **is like** wrestling in some ways. Touch is more important than sight. Erik **was like** his teammates in many ways—strong, well trained, mentally tough, and able to tolerate pain.	Pattern A: Noun 1 + *be* + *like* + Noun 2
Pattern B: For Erik, wrestling and mountain climbing **are alike** in some ways. Erik and his teammates **were alike** in many ways.	Pattern B: Noun 1 + Noun 2 + *be* + *alike*
Compare: a. Erik **looks like** an athlete. He's tall and strong. b. Erik **is like** his teammates. He has a lot of experience and training.	Use *look like* to describe physical appearance. Use *be like* to describe an internal characteristic.

EXERCISE **30** ABOUT YOU Work with a student from another country. Ask a question with the words given. Use *be like*. The other student will answer.

EXAMPLE   families in the U.S. / families in your native country

A: Are families in the U.S. like families in your native country?

B: No, they aren't. Families in my native country are very big. Family members live close to each other.

1. an English class in the U.S. / an English class in your native country
2. your house (or apartment) in the U.S. / your house (or apartment) in your native country
3. the weather in this city / the weather in your hometown
4. food in your country / American food
5. women's clothes in your native country / women's clothes in the U.S.
6. a college in your native country / a college in the U.S.
7. American teachers / teachers in your native country
8. American athletes / athletes in your native country

12.13 | Same or Different

We show that two things are the same (or not) by using *the same as*. We show that two things are different by using *different from*.

Examples	Explanation
Pattern A: Football is not **the same as** soccer. Football is **different from** soccer.	Pattern A: Noun 1 is *the same as* Noun 2. Noun 1 is *different from* Noun 2.
Pattern B: Football and soccer are not **the same**. Football and soccer are **different**.	Pattern B: Noun 1 and Noun 2 are *the same*. Noun 1 and Noun 2 are *different*.
Language Note: You will hear some Americans say *different than*.	

EXERCISE **31** Tell if the two items are the same or different.

EXAMPLES boxing, wrestling

Boxing and wrestling are different.

fall, autumn

Fall is the same as autumn.

1. Michael Jordan, Michael Schumacher
2. baseball in Cuba, baseball in the U.S.
3. the Chicago Bulls, the Chicago Bears
4. a kilometer, 1,000 meters

5. L.A., Los Angeles
6. a mile, a kilometer
7. football, rugby
8. football rules, soccer rules

EXERCISE **32** *Combination Exercise.* Fill in the blanks in the following conversation.

A: I heard that you have a twin brother.

B: Yes, I do.

A: Do you and your brother look _____alike_____?
                                   (example)

B: No. He _____ look _____ me at all.
              (1)   *(2)*

A: But you're twins.

B: We're fraternal twins. That's different _____ identical
                                              (3)

twins who have the _____ genetic code. We're just
                        (4)

brothers who were born at _____ time. We're not
                                (5)

even the same _____ . I'm much taller than he is.
                    (6)

A: But you're _____ in many ways, aren't you?
                   (7)

B: No. We're completely _____ . I'm athletic and I'm on
                              (8)

the high school football team, but David hates sports. He's a much
_____ student than I am. He's much more
       (9)

_____ our mother, who loves to read and learn
       (10)

new things, and I _____ our father, who's athletic
                       (11)

and loves to build things.

A: What about your character?

B: I'm outgoing and he's very shy. Also we don't dress _____
                                                            (12)

at all.

He likes to wear neat, conservative clothes, but I prefer torn jeans and T-shirts.

A: From your description, it _____ like you're not even from
(13)

the same family.

B: We have one thing in common. We were both interested in

_____ girl at school. We both asked her out, but she
(14)

didn't want to go out with either one of us!

EXERCISE 33 *Combination Exercise.* This is a conversation between two women. Fill in the blanks with an appropriate word to complete the comparisons.

A: In the winter months, my husband doesn't pay as _____much_____
(example)

attention to me _____ he does to his football games.
(1)

B: Many women have the same problem _____ you do.
(2)

These women are called football "widows" because they lose their husbands during football season.

A: I feel _____ a widow. My husband is in front of the TV
(3)

all day on the weekends. In addition to the football games, there are pre-game shows. These shows last _____ long as the
(4)

game itself.

B: I know what you mean. He's no different _____ my
(5)

husband. During football season, my husband is _____
(6)

interested in watching TV _____ he is in me. He looks
(7)

_____ a robot sitting in front of the TV. When I complain, he
(8)

tells me to sit down and join him.

A: It sounds _____ all men act _____ during football
(9)   (10)

season.

B: To tell the truth, I don't like football at all.

A: I don't either. I think soccer is much _____ interesting than
(11)

football.

B: Soccer is very different _____ football. I think the action is
(12)

_____ exciting. And it's more fun to watch the foot work of the
(13)

soccer players. Football players look _____ big monsters with
(14)

their helmets and padded shoulders. They don't look handsome at all.

A: Soccer is not _____ popular in the U.S. _____
(15)   (16)

it is in other countries. I wonder why.

B: What's your favorite team?

A: I like the Chicago Fire.

B: In my opinion they're not _____ good as the Los Angeles
(17)

Galaxy. But to tell the truth, I'm not very interested in sports at all.
When our husbands start watching football next season, let's do our
favorite sport: shopping. We can spend _____ time shopping
(18)

as they spend in front of the TV.

A: I was just thinking the same thing! You and I think _____.
(19)

Instead of being football widows, they can be shopping "widowers."

SUMMARY OF LESSON 12

1. Simple, Comparative, and Superlative Forms
   SHORT WORDS
   >Jacob is **tall.**
   >Mark is **taller than** Jacob.
   >Bart is **the tallest** member of the basketball team.

   LONG WORDS
   >Golf is **popular** in the U.S.
   >Baseball is **more popular than** golf.
   >Soccer is **the most popular** game in the world.

2. Other Kinds of Comparisons
   >She looks **as young as** her daughter.
   >She speaks English **as fluently** as her husband.
   >She is **the same age as** her husband.
   >She and her husband are **the same age.**
   >She works **as many hours as** her husband.
   >She doesn't have **as much time as** her husband.
   >She works **as much as** her husband.

3. Comparisons with *Like*

> She**'s like** her mother. (She and her mother **are alike.**) They're both athletic.
> She **looks like** her sister. (She and her sister **look alike.**) They're identical twins.
> Coke **tastes like** Pepsi. (They taste **alike.**)
> Western music doesn't **sound like** Asian music. (They don't **sound alike.**)

4. Comparisons with *Same* and *Different*

> Football is **different from** soccer.
> My uniform is **the same as** my teammates' uniforms.

EDITING ADVICE

1. Don't use a comparison word when there is no comparison.

   New York is a bigger city.

2. Don't use *more* and *-er* together.

   He is ~~more~~ older than his teacher.

3. Use *than* before the second item of comparison.

   than
   He is younger ~~that~~ his wife.

4. Use *the* before a superlative form.

   the
   The Nile is ∧ longest river in the world.

5. Use a plural noun in the phrase "one of the [superlative] [nouns]."

   cities
   Chicago is one of the biggest ~~city~~ in the U.S.

6. Use the correct word order.

   speaks more
   She ~~more speaks~~ than her husband.

   more time
   I have ~~time more~~ than you.

7. Use *be like* for similar character. Use *look like* for a physical similarity.

   s
   He ~~is~~ look ∧ like his brother. They both have blue eyes and dark hair.
   He is ~~look~~ like his sister. They are both talented musicians.

8. Don't use *the* and a possessive form together.

My ~~the~~ youngest son likes soccer.

9. Use the correct negative for *be like, look like, sound like, feel like,* etc.

don't
I'~~m not~~ look like my father.

does
He ~~is~~ not act like a professional athlete.

LESSON 12 TEST / REVIEW

PART **1**   Find the mistakes with the underlined words, and correct them. Not every sentence has a mistake. If the sentence is correct, write *C*.

EXAMPLES   She ~~is~~ look_slike her sister. They both have curly hair.

A house in the suburbs is <u>much more expensive</u> than a house in the city.   C

1. I am <u>the same</u> tall as my brother.
2. New York City is the <u>larger</u> city in the U.S.
3. That man is smarter <u>that</u> his wife.
4. The youngest student in the class has <u>more better</u> grades than you.
5. A big city has <u>crime more</u> than a small town.
6. I have three sons. My <u>oldest</u> son is married.
7. I visited many American cities, and I think that San Francisco is the <u>more</u> beautiful city in the U.S.
8. New York is one of the largest <u>city</u> in the world.
9. My uncle is <u>the most intelligent</u> person in my family.
10. She <u>faster types</u> than I do.
11. Texas is one of the biggest <u>state</u> in the U.S.
12. He <u>more carefully drives</u> than his wife.
13. Paul is one of the youngest <u>students</u> in this class.
14. She is richer than her best friend, but her friend is happier <u>than</u>.
15. <u>My the</u> best grade this semester was A–.
16. She <u>isn't look</u> like her sister at all. She's short and her sister is tall.

PART **2** Fill in the blanks.

EXAMPLE Pepsi is ____the same____ color ____as____ Coke.

1. She's 35 years old. Her husband is 35 years old. She and her husband are _____ age.

2. She earns $30,000 a year. Her husband earns $35,000. She doesn't earn as _____ her husband.

3. The little girl _____ like her mother. They both have brown eyes and curly black hair.

4. My name is Sophia Weiss. My teacher's name is Judy Weiss. We have _____ last name.

5. Chinese food is different _____ American food.

6. A dime isn't the same _____ a nickel. A dime is smaller.

7. She is as tall as her husband. They are the same _____

_____.

8. I drank Pepsi and Coke, and I don't know which is which. They have the same flavor. To me, Pepsi _____ like Coke.

9. She _____ like her husband in many ways. They're both intelligent and hard-working. They both like sports.

10. **A:** Are you like your mother?

B: Oh, no. We're not _____ at all! We're completely different.

11. Please finish this test _____ possible!

12. *Borrow* and *lend* don't have _____ meaning. *Borrow* means take. *Lend* means give.

13. My two sisters look _____. In fact, some people think they're twins.

Classroom Activities

1. Work with a partner. Find some differences between the two of you. Then write five sentences that compare you and your partner. Share your answers in a small group or with the whole class.

   EXAMPLES   I'm taller than Alex.

   Alex is taking more classes than I am.

2. Form a small group (about 3–5 people) with students from different native countries, if possible. Make comparisons about your native countries. Include a superlative statement. (If all the students in your class are from the same native country, compare cities in your native country.)

   EXAMPLES   Cuba is closer to the U.S. than Peru is.

   China has the largest population.

   Cuba doesn't have as many resources as China.

3. Work with a partner. Choose one of the categories below, and compare two examples from this category. Use any type of comparative method. Write four sentences. Share your answers with the class.

a. countries	e. cities	i. sports
b. cars	f. animals	j. athletes
c. restaurants	g. types of transportation	
d. teachers	h. schools	

   EXAMPLE   animals

   A dog is different from a cat in many ways.

   A dog can't jump as high as a cat.

   A dog is a better pet than a cat, in my opinion.

   A cat is not as friendly as a dog.

4. Compare the U.S. to another country you know. Tell if the statement is true in the U.S. or in the other country. Form a small group and explain your answers to the others in the group.

	Country	The U.S.
People have more free time.		
People have more political freedom.		
Families are smaller.		
Children are more polite.		
Teenagers have more freedom.		
People are friendlier.		
The government is more stable.		
Health care is better.		
There is more crime.		
There are more poor people.		
People are generally happier.		
People are more open about their problems.		
Friendship is more important.		
Women have more freedom.		
Schools are better.		
Job opportunities are better.		
Athletes make more money.		
Children have more fun.		
People dress more stylishly.		
Families are closer.		
People are healthier.		

5. Game—Test your knowledge of world facts.

Form a small group. Answer the questions below with other group members. When you're finished, check your answers. (Answers are at the bottom of the next page.) Which group in the class has the most correct answers?

1. Which athlete said, "I'm the greatest"?

    Michael Jordan   Pelé   Muhammad Ali   Serena   Williams

2. Where is the tallest building in the world?

    New York City   Chicago   Tokyo   Taipei

3. What country has the largest population?

the U.S.   India   China   Russia

4. Which country has the largest area?

the U.S.   China   Canada   Russia

5. What is the tallest mountain in the world?

Mount McKinley   Mount Everest

Mount Kanchenjunga   Mount Lhotse

6. Which state in the U.S. has the smallest population?

Alaska   Wyoming   Rhode Island   Vermont

7. What is the longest river in the world?

the Mississippi   the Missouri   the Nile   the Amazon

8. What is the biggest animal?

the elephant   the rhinoceros   the giraffe   the whale

9. What is the world's largest island?

Greenland   New Guinea   Borneo   Madagascar

10. What country has the most time zones?

China   Russia   the U.S.   Canada

11. What is the world's largest lake?

Lake Superior   Lake Victoria

the Caspian Sea   the Aral Sea

12. Which planet is the closest to the Earth?

Mercury   Venus   Mars   Saturn

13. Where is the world's busiest airport?

Chicago   New York   Los Angeles   London

14. Which is the most popular magazine in the U.S.?

Time   *Sports Illustrated*   *TV Guide*   *People Weekly*

15. What language has the largest number of speakers?

English   Chinese   Spanish   Russian

16. Which country has the most neighboring countries?

China   Russia   Saudi Arabia   Brazil

The answers are: 1. Ali 2. Taipei 3. China 4. Russia 5. Mount Everest 6. Wyoming 7. the Nile 8. the whale 9. Greenland 10. Russia 11. the Caspian Sea 12. Mars 13. Chicago 14. *TV Guide* 15. Chinese 16. China (It has 16 neighboring countries.)

6. Look at the list of jobs below. Use the superlative form to name a job that matches each description. You may discuss your answers in a small group or with the entire class.

EXAMPLE   interesting
          In my opinion, a psychologist has the most interesting job.

coach	referee
psychologist	letter carrier
computer programmer	athlete
high school teacher	actress
factory worker	photojournalist
doctor	firefighter
police officer	politician
engineer	nurse

(you may add other professions)

a. interesting _____

b. dangerous _____

c. easy _____

d. tiring _____

e. dirty _____

f. boring _____

g. exciting _____

h. important _____

i. challenging _____

j. difficult _____

Write a short composition comparing one of the sets of items below:

- two stores where you shop for groceries
- watching a movie at home and at a movie theater
- two friends of yours
- you and your parents
- football and soccer (or any two sports)
- clothing styles in the U.S. and your native country
- life in the U.S. (in general) and life in your native country
- your life in the U.S. and your life in your native country
- the American political system and the political system in your native country
- schools (including teachers, students, classes, etc.) in the U.S. and schools in your native country
- American families and families in your native country

Talk About it

1. Do athletes in other countries make a lot of money?

2. Do children in most countries participate in sports? Which sports?

Outside Activity

Interview someone who was born in the U.S. Get his or her opinion about the superlative of each of the following items. Share your findings with the class.

- prestigious job in the U.S.
- beautiful city in the U.S.
- popular TV program
- terrible tragedy in American history
- big problem in the U.S.
- handsome or beautiful actor
- good athlete
- good sports team

Internet Activities

1. Find an article about Michael Jordan on the Internet. Print the article and circle some interesting facts.

2. Find an article about an athlete that you admire. Print the article and circle some interesting facts.

3. Find an article about Enrique Oliu. Summarize the article. What makes Oliu so special?

4. Visit the Olympics Web site or a Web site with sports statistics and information. Find out which country has won the most medals in a particular sport. Which sport is the newest to be an Olympic event? Which athlete has the most Olympic medals?

Additional Activities at www.http://elt.thomson.com/gic

GRAMMAR

Passive Voice and Active Voice

CONTEXT: The Law

Jury Duty

Unusual Lawsuits

13.1 | The passive Voice and the Active Voice—An Overview

	Examples			Explanation
Active	Subject	Active Verb	Object	The **active voice** focuses on the person who performs the action. The subject is active.
	The thief	**stole**	the bicycle.	
	The police	**arrested**	the thief. 	
Passive	Subject	Passive Verb	*By* Phrase	The **passive voice** focuses on the receiver or the result of the action. The subject is passive. The person who does the action is in the *by* phrase.
	The bicycle	**was stolen**	by the thief.	
	The thief	**was arrested**	by the police.	
Passive	Subject	Passive Verb		Many passive sentences do not contain a *by* phrase.
	The thief	**was taken**	to jail.	
	The bicycle	**will be returned**.		

JURY DUTY

Before You Read

1. Have you ever been to court?

2. Have you ever seen a courtroom in a movie or TV show?

All Americans **are protected** by the Constitution. No one person can decide if a person is guilty of a crime. Every citizen has the right to a trial by jury. When a person **is charged** with a crime,   he **is considered** innocent until the jury decides he is guilty.

Most American citizens **are chosen** for jury duty at some time in their lives. How **are** jurors **chosen?**   The court gets the names of citizens from lists of taxpayers, licensed drivers, and voters. Many people **are called** to the courthouse for the selection of a jury. From this large number, 12 people **are chosen**. The lawyers and the judge ask each person questions to see if the person is going to be fair. If the person has made any judgment about the case before hearing the facts presented in the trial, he **is** not **selected**. If the juror doesn't understand enough English, he **is** not **selected**. The court needs jurors who can understand the facts and be open-minded. When the final jury selection **is made**, the jurors must raise their right hands and promise to be fair in deciding the case.

Sometimes a trial goes on for several days or more. Jurors **are** not **permitted** to talk with family members and friends about the case. In some cases, jurors **are** not **permitted** to go home until the case is over.   They stay in a hotel and **are** not **permitted** to watch TV or read newspapers that give information about the case.

After the jurors hear the case, they have to make a decision. They go to a separate room and talk about what they heard and saw in the courtroom. When they are finished discussing the case, they take a vote.

Jurors **are paid** for their work. They receive a small amount of money per day. Employers must give a worker permission to be on a jury. Being on a jury **is considered** a very serious job.

13.2 | The Passive Voice

Examples			Explanation
	Be	Past Participle	The passive voice uses a form of *be* (any tense) + the past participle.
The jurors	**are**	**chosen** from lists.	
My sister	**was**	**selected** to be on a jury.	
The jurors	**will be**	**paid** for jury duty.	
Compare Active (A) and Passive (P): (A) Ms. Smith *paid* her employees at the end of the week. (P) Ms. Smith *was paid* for being a juror.			The verb in active voice (A) shows that the subject (Ms. Smith) performed the action of the verb. The verb in passive voice (P) shows that the subject (Ms. Smith) did not perform the action of the verb.
I was helped **by the lawyer.** My sister was helped **by him** too.			When a performer is included after a passive verb, use *by* + noun or object pronoun.

Read the following sentences. Decide if the underlined verb is active *(A)* or passive *(P)*.

EXAMPLES   I <u>received</u> a letter from the court.   A

I <u>was told</u> to go to court on May 10.   P

1. The jury <u>voted</u> at the end of the trial.
2. The jurors <u>received</u> $20 a day.
3. Some jurors <u>were told</u> to go home.
4. Not every juror <u>will be needed</u>.
5. Twelve people <u>were selected</u> for the jury.
6. The judge <u>told</u> the jurors about their responsibilities.
7. My sister <u>has been selected</u> for jury duty three times.
8. You <u>will be paid</u> for jury duty.
9. A juror <u>must be</u> at least 18 years old and an American citizen.
10. The judge and the lawyers <u>ask</u> a lot of questions.

13.3 | Passive Voice—Form and Use

Form: The passive voice can be used with different tenses and with modals. The tense of the sentence is shown by the verb *be*. Use the past participle with every tense.

Tense	Active	Passive (*Be* + Past Participle)
Simple Present	They **take** a vote.	A vote **is taken**.
Simple Past	They **took** a vote.	A vote **was taken**.
Future	They **will take** a vote. They **are going to take** a vote.	A vote **will be taken**. A vote **is going to be taken**.
Present Perfect	They **have taken** a vote.	A vote **has been taken**.
Modal	They **must take** a vote.	A vote **must be taken**.

Language Notes:

1. An adverb can be placed between the auxiliary verb and the main verb.

   The jurors **are *always* paid**.

   Noncitizens **are *never* selected** for jury duty.

2. If two verbs in the passive voice are connected with *and*, do not repeat *be*.

   The jurors **are taken** to a room and **shown** a film about the court system.

Use: The passive voice is used more frequently without a performer than with a performer.	
Examples	**Explanation**
English **is spoken** in the U.S. Independence Day **is celebrated** in July.	The passive voice is used when the action is done by people in general.
The jurors **are given** a lunch break. The jurors **will be paid** at the end of the day. Jurors **are** not **permitted** to talk with family members about the case.	The passive voice is used when the actual person who performs the action is of little or no importance.
a. The criminal **was arrested**. b. The students **will be given** a test on the passive voice.	The passive voice is used when it is obvious who performed the action. In (a), it is obvious that the police arrested the criminal. In (b), it is obvious that the teacher will give a test.
Active: The lawyers **presented** the case yesterday. Passive: The case **was presented** in two hours. Active: The judge and the lawyers **choose** jurors. Passive: People who don't understand English **are not chosen.**	The passive voice is used to shift the emphasis from the performer to the receiver of the action.

EXERCISE **2** Change to the passive voice. (Do not include a *by* phrase.)

	ACTIVE	PASSIVE
EXAMPLE	They chose him.	He was chosen.
1.	They will choose him.	
2.	They always choose him.	
3.	They can't choose him.	
4.	They have never chosen us.	
5.	They didn't choose her.	
6.	They shouldn't choose her.	

EXERCISE **3** Fill in the blanks with the passive voice of the verb in parentheses
( ). Use the present tense.

EXAMPLE Jurors _____are chosen_____ from lists.
                    (choose)

1. Only people over 18 years old _____ for jury
                                           (select)
   duty.

2. Questionnaires _____ to American citizens.
                          (send)

3. The questionnaire _____ out and
                              (fill)

   _____.
           (return)

4. Many people _____ to the courthouse.
                       (call)

5. Not everyone _____.
                       (choose)

6. The jurors _____ a lot of questions.
                     (ask)

7. Jurors _____ to discuss the case with
                 (not/permit)
   outsiders.

8. Jurors _____ a paycheck at the end of the
                 (give)
   day for their work.

EXERCISE **4** Fill in the blanks with the passive voice of the verb in parentheses
( ). Use the past tense.

EXAMPLE I _____was sent_____ a letter.
                  (send)

1. I _____ to go to the courthouse on Fifth Street.
         (tell)

2. My name _____.
                   (call)

3. I _____ a form to fill out.
             (give)

4. A video about jury duty _____ on a large TV.
                                    (show)

5. The jurors _____ to the third floor of the building.
                     (take)

6. I _____ a lot of questions by the lawyers.
             (ask)

7. I _____.
         (not / choose)

8. I _____ home before noon.
             (send)

EXERCISE **5**   Fill in the blanks with the passive voice of the verb in parentheses
( ). Use the present perfect tense.

EXAMPLE   The jurors _____*have been given*_____ a lot of information.
                              (give)

1. Many articles _____ about the courts.
                                    (write)

2. Many movies _____ about the courts.
                                    (make)

3. Many people _____ for jury duty.
                                    (choose)

4. Your name _____ for jury duty.
                                    (select)

5. The jurors _____ for their work.
                                    (pay)

6. The check _____ at the door.
                                    (leave)

7. The money _____ in an envelope.
                                    (put)

EXERCISE **6**   The people called to jury duty are getting instructions about what
to expect. Fill in the blanks with the passive voice of the verb in
parentheses ( ). Use the future tense.

EXAMPLE   You _____*will be taken*_____ to a courtroom.
                              (take)

1. You _____ to stand up when the
                    (tell)
judge enters the room.

2. Each of you _____ a lot of questions.
                                (ask)

3. The lawyers _____ .
                                (introduce)

4. Information about the case _____ to
                                        (give)
you.

5. You _____ to eat in the courtroom.
                (not/allow)

6. Twelve of you _____ .
      (select)

7. If you do not speak and understand English well, you

   _____ .
            (not / pick)

8. Besides the 12 jurors, two alternates[1]

   _____ .
            (choose)

9. The rest of you _____ home.
                        (send)

10. All of you _____ .
                    (pay)

EXERCISE **7**   Fill in the blanks with the passive voice of the underlined verbs.
Use the same tense.

EXAMPLE   The jury <u>took</u> a vote. The vote ____was taken____ after three hours.

1. The lawyers <u>asked</u> a lot of questions. The questions

   _____ to find facts.

2. The court <u>will pay</u> us. We _____ $20 a
   day.

3. They <u>told</u> us to wait. We _____ to wait
   on the second floor.

4. They <u>gave</u> us instructions. We _____
   information about the law.

5. People <u>pay</u> for the services of a lawyer. Lawyers

   _____ a lot of money for their services.

6. You <u>should use</u> a pen to fill out the form. A pen

   _____ for all legal documents.

7. They <u>showed</u> us a film about the court system. We

   _____ the film before we went to the

   courtroom.

[1] An *alternate* takes the place of a juror who cannot serve for some reason (such as illness).

13.4 | Negatives and Questions with Passive Voice

Compare affirmative statements to negative statements and questions with the passive voice.

Simple Past	Present Perfect
The jurors **were paid.**	I **have been chosen** for jury duty several times.
They **weren't paid** a lot.	I **haven't been chosen** this year.
Were they **paid** in cash?	**Have** you ever **been chosen?**
No, they **weren't.**	No, I **haven't.**
How much **were** they **paid?**	How many times **have** you **been chosen?**
Why **weren't** they **paid** in cash?	Why **haven't** you **been chosen?**

Language Note:
Never use *do, does,* or *did* with the passive voice.
   Wrong: The juror **didn't** paid.

EXERCISE **8**   Change to the negative form of the underlined words.

EXAMPLE   I <u>was selected</u> for jury duty last year. I ___wasn't selected___ this year.

1. The jurors <u>are paid</u>. They _____ a lot of money.

2. Twelve people <u>were chosen</u>. People who don't speak English well

   _____.

3. Jurors <u>are allowed</u> to talk with other jurors about the case. They

   _____ to talk to friends and family about

   the case.

4. We <u>were told</u> to keep an open mind. We _____

   how to vote.

5. We <u>have been given</u> instructions. We _____

   our checks yet.

EXERCISE **9**   Change the statements to questions using the words in parentheses
( ).

EXAMPLE   The jurors are paid. (how much)
   <u>How much are the jurors paid?</u>

1. Some people aren't selected. (why)

   _____

2. The jurors are given a lunch break. (when)

   _____

3. I wasn't chosen for the jury. (why)

_____

4. You were given information about the case. (what kind of information)

_____

5. A film will be shown. (when)

_____

6. Several jurors have been sent home. (why)

_____

7. The jurors should be paid more money. (why)

_____

8. We were told to go to the courtroom. (when)

_____

9. The jury has been instructed by the judge. (why)

_____

UNUSUAL LAWSUITS

Before You Read

1. Are drivers permitted to use cell phones in the area where you live?

2. Have you read about any unusual court cases in the newspaper or heard about any on TV?

Read the following article. Pay special attention to the active and passive voice.

When a person is **injured** or **harmed**, it is the court's job to determine who is at fault. Most of these cases never **make** the news. But a few of them **appear** in the newspapers and on the evening news because they are so unusual.

In 1992, a fast-food restaurant **was sued** by a 79-year-old woman in New Mexico who **spilled** hot coffee on herself while driving. She **suffered** third-degree burns on her body. At first the woman **asked** for $11,000 to cover her medical expenses. When the restaurant **refused**, the case **went** to court and the woman **was awarded** nearly $3 million dollars.

In 2002, a group of teenagers **sued** several fast-food chains for serving food that **made** them fat. The case **was thrown** out of court. According to Congressman Ric Keller, Americans **have to** "get away from this new culture where people always try to play the victim and blame others for their problems." Mr. Keller, who is overweight and **eats** at fast-food chains once every two weeks, **said** that suing "the food industry is not going to make a single individual any skinnier. It **will** only **make** the trial attorneys' bank accounts fatter."

In June 2004 an Indiana woman **sued** a cell phone company for causing an auto accident in which she was involved. The court **decided** that the manufacturer of a cell phone cannot **be held** responsible for an auto accident involving a driver using its product. In March 2000, a teenage girl in Virginia was **struck** and **killed** by a driver conducting business on a cell phone. The girl's family **sued** the driver's employer for $30 million for wrongful death. They said that it was the company's fault because employees **are expected** to conduct business while driving. The family **lost** its case.

We **are protected** by the law. But as individuals we **need to take** personal responsibility and not blame others for our mistakes. The court system **is designed** to protect us; it **is** up to us to make sure that trials remain serious.

Source: The Cellular Telecommunications & Internet Association

13.5 | Choosing Active Voice or Passive Voice

Examples	Explanation
(A) A driver using a cell phone **caused** the accident. (P) The accident **was caused** by a driver using a cell phone. (A) A driver **struck** and **killed** a teenager. (P) A teenager **was struck** and **killed** by a driver.	When the sentence has a specific performer, we can use either the active (A) or passive(P) voice. The active voice puts more emphasis on the person who performs the action. The passive voice puts more emphasis on the action or the result. The performer is mentioned in a *by* phrase (*by the driver, by a woman, by the court*). The active voice is more common than the passive voice when there is a specific performer.
(P) The obesity case **was thrown** out of court. (P) The manufacturer of a cell phone **cannot be held** responsible for a car accident. (P) Some employees **are expected** to conduct business while driving.	When there is no specific performer or the performer is obvious, the passive voice is usually used.
(P) It **was found** that six percent of accidents are the result of driver distraction. (P) It **is believed** that cell phone use distracts drivers.	Often the passive voice is used after *it* when talking about findings, discoveries, or general beliefs.
(A) The woman **went** to court. (A) The accident **happened** in Virginia. (A) Unusual court cases **appear** in the newspaper. (A) The teenager **died.**	Some verbs have no object. We cannot make these verbs passive. Some verbs with no object are: happen go fall become live sleep come look die seem work recover be remain arrive stay appear seem run sound grow depend wake up leave (a place)
(A) **She** sued **them.** (p) **They** were sued by **her.** (A) **He** helps **us.** (P) **We** are helped by **him.**	Notice the difference in pronouns in an active sentence and a passive sentence. After *by*, the object pronoun is used.

Language Note:
Even though *have* and *want* are followed by an object, these verbs are not usually used in the passive voice.
> He **has** a cell phone. (*Not:* A cell phone is had by him.)
> She **wants** a new car. (*Not:* A new car is wanted by her.)

EXERCISE **10**   Change these sentences from active to passive voice.   Mention the performer in a *by* phrase. Use the same tense.

EXAMPLE   An Indiana woman <u>sued</u> the cell phone company.
The cell phone company was sued by an Indiana woman.

1. Employees <u>use</u> cell phones.

2. A driver <u>hit</u> a pedestrian.

3. The court <u>threw out</u> the case.

4. Distracted drivers <u>cause</u> accidents.

5. Congress <u>makes</u> the laws.

6. <u>Should</u> the government <u>control</u> cell phone use?

7. The president <u>signs</u> a new law.

8. The court <u>has decided</u> the case.

9. The judge <u>will make</u> a decision.

10. Fast-food restaurants <u>sell</u> hamburgers and fries.

EXERCISE **11**   The following sentences would be better in passive voice without a performer. Change them. Use the same tense.

EXAMPLE   They <u>paid</u> me for jury duty.
I was paid for jury duty.

1. They <u>sent</u> me a questionnaire.

2. They <u>have taken</u> us to a separate room.

3. They <u>told</u> us not to discuss the case.

4. They <u>will choose</u> 12 people.

_____

5. <u>Has</u> someone <u>selected</u> your name?

_____

6. They <u>didn't permit</u> us to read any newspapers.

_____

7. They <u>will not select</u> him again for jury duty.

_____

8. <u>Will</u> they <u>pay</u> you?

_____

9. They <u>don't allow</u> us to eat in the courtroom.

_____

10. Someone <u>has called</u> my name.

_____

EXERCISE **12**   The following sentences would be better in active voice. Change them to active voice. Use the same tense.

EXAMPLE   **Fast food <u>is eaten</u> by Mr. Keller.**

Mr. Keller eats fast food.

1. A cell phone <u>was had</u> by the driver.

_____

2. Hot coffee <u>was spilled</u> by the driver.

_____

3. <u>Is</u> a cell phone <u>used</u> by you?

_____

4. The car <u>has been driven</u> by me.

_____

5. A lot of money <u>is made</u> by lawyers.

_____

6. A headset <u>should be used</u> by drivers with cell phones.

_____

7. Business <u>is conducted</u> by me from my car.

_____

8. The news <u>is watched</u> by us every night.

_____

9. Fast food <u>is eaten</u> by a lot of teenagers.

_____

10. The accident <u>will be reported</u> by them.

_____

EXERCISE 13 Fill in the blanks with the active or passive voice of the verb in parentheses ( ).Use the tense or modal given.

In about 40 countries, laws _____<u>have been passed</u>_____ that
(*example:* present perfect: *pass*)
prohibit drivers from using cell phones. In the U.S., the law _____
_____ on the place where you
(*1 present: depend*)
_____. In New York, for example, the use of
(*2 present: live*)
hand-held cell phones while driving _____, but
(*3 present: prohibit*)
the use of hands-free units _____. A driver who
(*4 present: permit*)
_____ this law can be fined $100 for a first
(*5 present: not/obey*)
offense, $200 for a second, and $500 after that. Other states
_____ to become tougher on drivers who use
(*6 present perfect: start*)
cell phones.

However, even when drivers _____
(*7 present: use*)
hands-free cell phones, they still _____ accidents.
(*8 present: cause*)
Drivers _____ their hands off the wheel to make
(*9 must/take*)
or end a call. The problem _____ if drivers
(*10 can/reduce*)
_____ voice-activated cell phones.
(*11 present: use*)
But the problem of driver distraction is not only a result of cell phones. According to one study conducted, it was found that six percent of accidents _____ by drivers who are not
(*12 present: cause*)
paying attention. But the distractions were not just from cell phones.
This study _____ that drivers
(*13 past: determine*)
_____ by many things: eating, putting on makeup,
(*14 present distract*)

reading, reaching for things, changing stations on the radio—as well as by cell phone use. It is clear that all drivers _____

(15 present: need)

to give driving their full attention.

EXERCISE 14   Fill in the blanks with the passive or active voice of the verb in parentheses ( ), using the past tense.

A: Why weren't you at work last week? Were you sick?

B: No. I _____ was chosen _____ to be on a jury.

                (example: choose)

A: How was it?

B: It was very interesting. A man _____ for fighting

                        (1 arrest)

   with a police officer.

A: Oh. How was the jury selection process?

B: The jury selection was interesting too. But it took half a day to choose 12 people.

A: Why?

B: The judge and lawyers _____ more than 50 people.

                      (2 interview)

A: Why so many people?

B: Well, several people _____ the judge's

                  (3 not/understand)

   questions. They _____ English very well. And a

             (4 not/speak)

   woman _____ the judge that she was very sick.

          (5 tell)

   The judge _____ her permission to leave. I don't

          (6 give)

   know why the other people _____.

                  (7 not/choose)

A: What kind of questions _____ by the

                (8 you/ask)

   judge and lawyers?

B: First the lawyers _____ to see if we could be fair.

           (9 want)

   Some jurors _____ that they had a bad experience

          (10 say)

   with a police officer. Those jurors _____.

                  (11 not/select)

A: Why not?

B: Because the judge probably thought they couldn't be fair in this case.

A: How long did the trial last?

B: Only two days.

A: _____ about the case with your
$\qquad$ *(12 you/talk)*

family when you _____ home the first night?
$\qquad$ *(13 go)*

B: Oh, no. We _____ not to talk to anyone about the
$\qquad$ *(14 tell)*

case. When it was over, I _____ my wife and kids
$\qquad$ *(15 tell)*

about it.

A: How long did it take the jurors to make a decision?

B: About two hours. One of the jurors _____
$\qquad$ *(16 not/agree)*

with the other 11 jurors. We _____ about the
$\qquad$ *(17 talk)*

evidence until she changed her mind.

A: _____ you for the days you missed work?
(18 your boss/pay)

B: Of course. He had to pay me. That's the law.

A: Now that you've done it once, you won't have to do it again. Right?

B: That's not true. This was the second time I _____.
$\qquad$ *(19 choose)*

SUMMARY OF LESSON 13

1. Active and Passive Voice

Active	Passive
He **drove** the car.	The car **was driven** by him.
He **didn't drive** the car.	The car **wasn't driven** by him.
He **will drive** the car.	The car **will be driven** by him.
He **has driven** the car.	The car **has been driven** by him.
He often **drives** the car.	The car **is** often **driven** by him.
He **should drive** the car.	The car **should be driven** by him.
Did he **drive** the car?	**Was** the car **driven** by him?
When **did** he **drive** the car?	When **was** the car **driven** by him?

2. The Active Voice

Examples	Explanation
I **bought** a new cell phone. He **eats** fast food. We **will drive** the car.	In most cases, the active voice is used when there is a choice between active and passive.
The accident **happened** last month. She **went** to court.	When there is no object, the active voice must be used. There is no choice.

3. The Passive Voice

Examples	Explanation
I **was chosen** for jury duty. My cell phone **was made** in Japan.	Use the passive voice when the performer is not known or is not important.
The criminal **was taken** to jail. Some employees **are expected** to conduct business while driving.	Use the passive voice when the performer is obvious.
Cell phones **are used** all over the world. Jury duty **is considered** a responsibility of every citizen.	Use the passive voice when the performer is everybody or people in general.
The court paid me. I **was paid** at the end of the day. The coffee was very hot. The coffee **was bought** at a fast-food restaurant.	shifted from the performer to the receiver of the action.
It **was discovered** that many accidents are the result of driver distraction. It **is believed** that a person can get a fair trial in the U.S.	Use the passive voice with *it*, when talking about findings, discoveries, or beliefs.
Accidents **are caused** by distracted drivers. A fast-food company **was sued** by a woman in New Mexico.	Use the passive voice when we want to emphasize the receiver of the action more than the performer. (In this case, the performer is included in a *by* phrase.)

EDITING ADVICE

1. Never use *do, does,* or *did* with the passive voice.

   wasn't found
   The money ~~didn't find~~.

   were
   Where ~~did~~ the jurors taken?

2. Don't use the passive voice with *happen, die, sleep, work, live, fall,* or *seem.*

   My grandfather ~~was~~ died four years ago.

3. Don't confuse the *-ing* form with the past participle.

   taken
   The criminal was ~~taking~~ to jail.

4. Don't forget the *-ed* ending for a regular past participle.

   ed
   My cousin was select ∧ to be on a jury.

5. Don't forget to use *be* with a passive sentence.

   were
   The books ∧ found on the floor by the janitor.

6. Use the correct word order with adverbs.

   I was told ⟨never⟩ about the problem.

LESSON 13 TEST/REVIEW

PART **1**   Find the mistakes with the underlined words, and correct them. Not every sentence has a mistake. If the sentence is correct, write *C*.

been
EXAMPLES   The same mistake <u>has ∧ made</u> many times.

  We <u>were told</u> not to say anything.   C

1. Children <u>should taught</u> good behavior.

2. Parents <u>should teach</u> children good behavior.

3. I <u>never was given</u> any information about the test.

4. 1 <u>have been had</u> my car for three years.

5. The driver <u>was given</u> a ticket for driving without a seatbelt.

6. Where <u>did</u> your gloves <u>find</u>?

7. They <u>were find</u> in the back seat of a taxi.

8. Something <u>was happened</u> to my bicycle.

9. This carpet <u>has been cleaned</u> many times.

10. The answers <u>don't written</u> in my book.

PART 2   Change sentences from active to passive voice. Do not mention the performer. (The performer is in parentheses.) Use the same tense as the underlined verb.

EXAMPLE   (Someone) <u>took</u> my dictionary.

         My dictionary was taken.

1. (People) <u>speak</u> English in the U.S.

_____

2. (You) <u>can use</u> a dictionary during the test.

_____

3. (The police) <u>took</u> the criminal to jail.

_____

4. (People) <u>have seen</u> the president on TV many times.

_____

5. (Someone) <u>will take</u> you to the courtroom.

_____

6. (Someone) <u>has broken</u> the mirror into small pieces.

_____

7. (People) <u>expect</u> you to learn English in the U.S.

_____

8. (They) <u>don't allow</u> cameras in the courtroom.

_____

PART 3   Change the sentences from passive to active voice. Use the same tense.

EXAMPLE You <u>were told</u> by me to bring your books.

         I told you to bring your books.

1. You <u>have been told</u> by the teacher to write a composition.

_____

2. Your phone bill <u>must be paid</u>.

_____

3. You <u>are not allowed</u> by the teacher to use your books during a test.

_____

4. The tests <u>will be returned</u> by the teacher.

_____

5. When <u>are</u> wedding gifts <u>opened</u> by the bride and groom?

_____

6. Your missing car <u>was not found</u> by the police.

_____

PART **4** Fill in the blanks with the passive or active form of the verb in parentheses ( ). Use an appropriate tense.

EXAMPLES The tests _____*will be returned*_____ tomorrow.
　　　　　　　　　　　　　(will/return)

　　　　　　　The teacher _____*will return*_____ the tests.
　　　　　　　　　　　　　(will/return)

1. My neighbor had a heart attack and _____ to the
　　　　　　　　　　　　　　　　　　　　(take)

hospital in an ambulance yesterday.

2. I _____ my neighbor in the hospital tomorrow.
　　(will/visit)

3. I _____ the movie *Star Wars* five times.
　　　　(see)

4. This movie _____ by millions of people.
　　　　　　　(see)

5. I _____ a lot of friends.
　　　(have)

6. I _____ many times by my friends.
　　　(help)

7. Ten people _____ in the fire last night.
　　　　　　(die)

8. Five people _____ by the fire department in
　　　　　　　(rescue)

yesterday's fire.

9. Her husband _____ home from work at six o'clock
　　　　　　　(come)

every day.

10. He _____ home by his coworker last night.
    (drive)

11. The answer to your question _____ by anyone.
    (not/know)

12. Even the teacher _____ the answer to your question.
    (not/know)

EXPANSION ACTIVITIES

Classroom Activities

1. Form a small group and talk about the legal system in another country. Use the chart below to get ideas.

   Country: _____

	Yes	No
People are treated fairly in court.		
Citizens are selected to be on a jury.		
People are represented by lawyers in court.		
Lawyers make a lot of money.		
Famous trials are shown on TV.		
Punishment is severe for certain crimes.		
The death penalty is used in some cases.		
The laws are fair.		

2. Form a small group and tell about how a holiday is celebrated in your native culture. Use the chart below to get ideas.

	Yes	No
Gifts are given.		
The house is cleaned.		
Special clothing is worn.		
The house is decorated with special symbols of the holiday.		
Special food is prepared.		
Stores and businesses are closed.		
Special programs are shown on TV.		
Candles are used.		

3. Form two groups. One group should make a presentation telling why cell phone use should be permitted in cars. One group should make a presentation telling why cell phone use should *not* be permitted in cars.

1. Would you like to be on a jury? Why or why not?

2. In a small group, discuss your impressions of the American legal system from what you've seen on TV, from what you've read, or from your own experience.

3. Do you think drivers who use cell phones while driving cause accidents?

4. What laws should be changed in the U.S.? What laws should be added?

5. Do you think fast-food restaurants are responsible for obesity in the U.S.?

1. Write about an experience you have had with the court system in the U.S. or your native country.

2. Write about a famous court case that you know of. Do you agree with the decision of the jury?

3. Write about the advantages of owning a cell phone.

1. Watch a court movie, such as *The Firm, Witness to the Prosecution, Inherit the Wind, A Time to Kill, To Kill a Mockingbird, Presumed Innocent, Twelve Angry Men, A Civil Action,* or *The Client.* Write about your impressions of the American court system after watching one of these movies.

2. Watch a court TV show, such as *People's Court* or *Judge Judy.* What do you think of the judges' decisions on these shows?

3. Ask an American if he or she has ever been selected for a jury. Ask him or her to tell you about this experience.

1. At a search engine, type in *Insurance Information Institute* and *cell phones.* Find some statistics about drivers who use cell phones. Bring the information to class. Is there anything that surprises you?

2. Look for information about one of these famous American trials:
   a. the O.J. Simpson trial
   b. the Leopold and Loeb trial
   c. the Sacco and Vanzetti trial
   d. the Amistad trials
   e. the Scopes trial
   f. the Rosenberg trial
   g. the Bruno Hauptmann trial

(continued)

Answer these questions about one of the trials:

- What was the defendant accused of?

- When did the trial take place?

- How long did the trial last?

- Was the defendant found guilty?

 Additional Activities at http: // elt.thomson.com/gic

LESSON

14

GRAMMAR

Articles
Other / Another
Indefinite Pronouns

CONTEXT: Money

Kids and Money
Changing the American Dollar
The High Cost of a College Education

14.1 | Articles—An Overview

Articles precede nouns and tell whether a noun is definite or indefinite.	
Examples	**Explanation**
Do you have **a credit card?** I bought **an old house.**	The indefinite articles are *a* and *an.*
It's a holiday today. **The banks** are closed. There are many poor people in **the world.**	The definite article is *the.*
Money is important for everyone. **Children** like to spend money.	Sometimes a noun is used without an article.

KIDS AND MONEY

Before You Read

1. Do you think parents should give money to their children? At what age?

2. Do you think teenagers should work while they're in high school?

Read the following article. Pay special attention to nouns and the articles that precede them. (Some nouns have no article.)

> **Kids** in the U.S. like to spend **money.** In 2001, kids between the ages of 12 and 19 spent an average of $104 a week. Much of today's **advertising** is directed at kids. When you go into **a store,** you often hear **toddlers**[1] who are just learning to talk saying to their parents, "Buy me **a toy.** Buy me **some candy.**" Some kids feel **gratitude** when they receive **a dollar** or **a toy** from

[1]A *toddler* is a child between the ages of one and three.

a grandparent. But some kids feel a sense of entitlement[2]. Even during **the** hard economic **times** of the early 1990s, sales of **soft drinks, designer blue jeans, fast food, sneakers, gum,** and **dolls** remained high. One factor in parents' **generosity** is **guilt.** As **parents** become busier in their **jobs,** they often feel guilty about not spending **time** with their **kids.** Often they deal with their **guilt** by giving their kids **money** and **gifts.**

To help children understand **the value** of **money, parents** often give their **children an allowance. The child's spending** is limited to **the money** he or she receives each week. How much should parents give **a child** as **an allowance?** Some parents give **the child a dollar** for each year of his or her age. **A five**-**year**-**old** would get five dollars. **A fifteen**-**year**-**old** would get fifteen dollars. Some parents pay their kids extra for **chores,** such as taking out **the garbage** or shoveling **snow.** Other parents believe kids should do chores as part of their family responsibilities.

When is **the right time** to start talking to **kids** about **money?** According to Nathan Dungan, **a financial expert,** the right time is as soon as **kids** can say, "I want." By **the time** they start **school,** they must know there are **limits.**

14.2 | The Indefinite Article—Classifying or Identifying the Subject

Examples	Explanation
A doll is **a toy**. A toddler is **a small child**. An allowance is **a weekly payment** to children. A penny is **a one**-**cent coin**. "Big" is **an adjective**. "Inflation" is **an economic term**.	After the verb *be*, we use the indefinite articles *a* or *an* + singular count noun to define or classify the subject of the sentence. We are telling who or what the subject is. Singular subject + *is* + *a (n)* + (adjective) + noun
Jeans are **popular clothes**. Teenagers are **youny adults**. Chores are **everyday jobs**.	When we classify or define a plural subject, we don't use an article. Plural subject + *are* + (adjective)+noun

Language Note:
 We can also use *be* in the past tense to give a definition.
    The Depression **was** a difficult time in American history.
    Abraham Lincoln **was** an American president.	

[2]A sense of *entitlement* is a feeling that you have the right to receive something.

EXERCISE **1** Define the following words. Answers may vary.

EXAMPLE A toddler _is a small child._

1. A teenager _____
2. A quarter _____
3. A dime _____
4. A credit card _____
5. A wallet _____
6. Gold _____
7. Silver and gold _____

EXERCISE **2** Tell who these people are or were by classifying them. These people were mentioned in previous lessons in this book. Answers will vary.

EXAMPLE Martin Luther King, Jr. _was an African-American leader._

1. Albert Einstein _____
2. Michael Jordan _____
3. Erik Weihenmayer _____
4. Oprah Winfrey _____
5. George Dawson _____
6. Navajos _____
7. George Washington and Abraham Lincoln _____

14.3 | The Indefinite Article—Introducing a Noun

Examples	Explanation
She has **a son.** Her son has **a job.** Her son has **a checking account.**	Use *a* or *an* to introduce a singular noun.
He has (**some**) toys. He doesn't have (**any**) video games. Does he have (**any**) CDs?	Use *some* and *any* to introduce a plural noun. Use *any* for negatives and questions. *Some* and *any* can be omitted.
He has (**some**) money. He doesn't have (**any**) cash. Does he have (**any**) time?	Use *some* and *any* to introduce a noncount noun. Use *any* for negatives and questions. *Some* and *any* can be omitted.

EXERCISE **3**   Fill in the blanks with the correct word: *a, an, some,* or *any.*

EXAMPLE   There are _____ some _____ symbols on the back of a credit card.

1. Do you have _____ account with the bank?

2. Do you have _____ money in your savings account?

3. I have _____ twenty-dollar bill in my pocket.

4. I have _____ quarters in my pocket.

5. I have _____ money with me.

6. Do you have _____ credit cards?

7. I don't have _____ change.

8. Buy me _____ toy.

9. Buy me _____ candy.

10. I need _____ dollar.

11. Many teenagers want to have _____ job.

12. Does your little brother get _____ allowance?

EXERCISE **4**   A mother (M) and a son (S) are talking. Fill in the blanks with *a,*
an, some, or *any.*

S: I want to get ____ a ____ job.
                   (example)

M: But you're only 16 years old.

S: I'm old enough to work. I need to make _____ money.
                                              (1)

M: But we give you _____ allowance each week. Isn't
                        (2)

that enough money for you?

S: You only give me $15 a week. That's not even enough to buy

_____ CD or take _____ girl to
       (3)   *(4)*

_____ movie.
       (5)

M: If you work, what are you going to do about school? You won't have

_____ time to study. Do you know how hard it is to
       (6)

work and do well in school?

S: Of course, I do. You know I'm _____ good student.
                                         (7)

I'm sure I won't have _____ problems working part time.
                           (8)

M: Well, I'm worried about your grades falling. Maybe we should raise your allowance. That way you won't have to work.

S: I want to have my own money. I want to buy _____
(9)

new clothes. And I'm going to save money to buy _____
(10)

car someday.

M: Why do you want a car? You have _____ bike.
(11)

S: Bikes are great for exercise, but if my job is far away, I'll need a car for transportation.

M: So, you need _____ job to buy _____
(12)   (13)

car, and you need _____ car to get work.
(14)

S: Yes. You know, a lot of my friends work, and they're good students.

M: Well, let me think about it.

S: Mom, I'm not _____ baby anymore. I need
(15)

_____ job.
(16)

14.4 | The Definite Article

We use *the* to talk about a specific person or thing or a unique person or thing.

Examples	Explanation
The book talks about kids and money. **The author** wants to teach Kids to be responsible with money.	The sentences to the left refer to a specific object or person that is present. There is no other book or author present, so the listener knows which noun is referred to.
Many kids in **the world** are poor. **The first** chapter talks about small children. **The back** of the book has information about the author. When is **the right** time to talk to kids about money?	Sometimes there is only one of something. There is only one world, only one first chapter, only one back of a book. We use *the* with the following words: *first, second, next, last, only, same,* and *right.*
Where's **the** teacher? I have a question about **the** homework.	When students in the same class talk about **the** teacher, **the** textbook, **the** homework, **the** chalkboard, they are talking about a specific one that they share.
Did you read **the article about money?** Children often spend **the money they get from their grandparents.**	The sentences to the left refer to a specific noun that is defined in the phrase or clause after the noun: *the article **about money;** the money **they get from their grandparents.***

I'm going to **the** store after work. Do you need anything?
 The bank is closed. I'll go tomorrow.
 You should make an appointment with **the** doctor.
	We often use *the* with certain familiar places and people when we refer to the one that we usually use:
	     the bank the beach the bus
	     the zoo the post office the train
	     the park the doctor the movies
	     the store

a. I saw a **child** in the supermarket with her mother. b. **The child** kept saying, "Buy me this, buy me that." a. She used **a credit card.** b. She put **the credit card** back in her purse.	a. A noun is first introduced as an indefinite noun (with *a* or *an*). b. When referring to the same noun again, the definite article *the* is used.

My grandparents gave me lots of presents.
 Kim's kids have lots of toys.
	Don't use the definite article with a possessive form.
	     Wrong: My the grandparents
	     Wrong: Kim's the kids

EXERCISE **5** Fill in the blanks with the definite article *the*, the indefinite article *a* or *an*, or quantity words *any* or *some*.

Conversation 1: between two friends

A: Where are you going?

B: To _____the_____ bank. I want
        (example)

to deposit _____
                    (1)

check.

A: _____ bank is
        (2)

probably closed now.

B: No problem. I have

_____ ATM card.
        (3)

There's _____
                (4)

ATM on_____
                (5)

corner of Wilson and Sheridan.

A: I'll go with you. I want to get _____ cash.
                                            (6)

Later, at the ATM...

B: Oh,no. _____ ATM is out of order.
                (7)

A: Don't worry. There's _____ ATM in _____
                                (8)   *(9)*

supermarket near my house.

Conversation 2: between two students at the same school

A: Is there _____ cafeteria at this school?
                    (1)

B: Yes, there is. It's on _____ first floor of this building.
                                (2)

A: I want to buy _____ cup of coffee.
                        (3)

B: You don't have to go to _____ cafeteria. There's
                                    (4)

_____ coffee machine on this floor.
        (5)

A: I only have a one-dollar bill. Do you have _____ change?
                                                        (6)

B: There's _____ dollar-bill changer next to _____
                (7)   *(8)*

coffee machine.

Conversation 3: between two students (A and B) in the same class

A: Where's _____ teacher? It's already 7:00.
                (1)

B: Maybe she's absent today.

A: I'll go to _____ English office and ask if anyone
                    (2)

    knows where she is.

B: That's _____ good idea.
              (3)

A few minutes later...

A: I talked to _____ secretary in _____
              (4)   *(5)*

    English office. She said that _____ teacher just called.
                            (6)

    She's going to be about 15 minutes late. She had _____
                                    (7)

    problem with her car.

14.5 | Making Generalizations

When we make a generalization, we say that something is true of ALL members of a group.	
Examples	**Explanation**
a. **Children** like to copy their friends. b. **A child** likes to copy his or her friends. a. **Video games** are expensive. b. **A video game** is expensive.	There are two ways to make a generalization about a countable subject: a. Use *a* or *an* + singular noun OR b. Use no article + plural noun.
Money doesn't buy happiness. **Love** is more important than money. **Honesty** is a good quality.	To make a generalization about a noncount subject, don't use an article.
a. Children like **toys**. a. People like to use **credit cards**. b. Everyone needs **money**. b. No one has enough **time**.	Don't use an article to make a generalization about the object of the sentence. a. Use the plural form for count nouns. b. Noncount nouns are always singular.

Language Note:
Do not use *some* or *any* with generalizations.
Compare:
    I need **some money** to buy a new bike.
    Everyone needs **money**.

EXERCISE **6**   Decide if the statement is general (true of all examples of the subject), or specific (true of the pictures on this page or of specific objects that everyone in the class can agree on). Fill in the blanks with *a, an, the*, or Ø (for no article).

EXAMPLES   __Ø__ children like __Ø__ toys.

__The__ toys are broken.

1. _____ American teenager likes to have a job.

2. _____ teenager is shoveling snow to make money.

3. _____ teenagers like _____ cars.

4. _____ blue jeans are popular.

5. _____ blue jeans are torn.

6. _____ money is important for everyone.

7. _____ money on the table is mine.

8. Do you like _____ kids?

9. _____ American kids like to spend money.

10. _____ child is saying to her mother, "I want."

11. _____ coffee is hot.

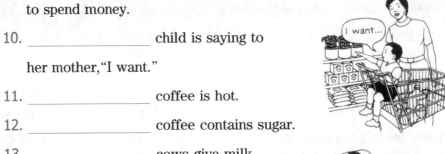

12. _____ coffee contains sugar.

13. _____ cows give milk.

14. _____ cows are eating grass.

15. _____ children are playing.

16. _____ children can generally learn a foreign language faster than adults.

EXERCISE **7** ABOUT YOU Tell if you like the following or not. For count nouns (C), use the plural form. For noncount nouns (NC), use the singular form.

EXAMPLES coffee (NC)   apple (C)
I like coffee.   I don't like apples.

1. tea (NC)   4. potato chip (C)   7. cookie (C)
2. corn (NC)   5. milk (NC)   8. pizza (NC)
3. peach (C)   6. orange (C)   9. potato (C)

EXERCISE **8** Fill in the blanks with *the, a, an, some, any,* or *Ø* (for no article). In some cases, more than one answer is possible.

A: Where are you going?

B: I'm going to __the__ post office. I need to buy _____ stamps.
    (example)   (1)

A: I'll go with you. I want to mail _____ package to my parents.
                                    (2)

B: What's in _____ package?
             (3)

A: _____ shirts for my father, _____ coat for my sister,
      (4)   (5)

   and _____ money for my mother.
         (6)

B: You should never send _____ money by mail.
                          (7)

A: I know. My mother never received _____ money that I sent in
                                     (8)

   my last letter. But what can I do? I don't have _____ checking
                                                     (9)

   account.

B: You can buy _____ money order at _____ bank.
                (10)   (11)

A: How much does it cost?

B: Well, if you have _____ account in _____ bank, it's
                      (12)   (13)

   usually free. If not, you'll probably have to pay a fee.

A: What about _____ currency exchange on Wright Street? Do
               (14)

   they sell _____ money orders?
              (15)

B: Yes.

A: Why don't we go there? We can save _____ time. It's on
                                       (16)

   _____ same street as _____ post office.
      (17)   (18)

EXERCISE **9**

Two women are talking. Fill in the blanks with *the, a, an, some, any,* or Ø (for no article). Answers may vary.

A: I bought my daughter _____a_____ new doll for her birthday. She's
        (example)

been asking me to buy it for her for two months. But she played with

_____ doll for about three days and then lost interest.
     (1)

B: That's how _____ kids are. They don't understand
                        (2)

_____ value of money.
     (3)

A: You're right. They think that _____ money grows on
                                          (4)

_____ trees.
     (5)

B: I suppose it's our fault. We have to set _____ good example. We
                                                    (6)

buy a lot of things we don't really need. We use _____ credit
                                                         (7)

cards instead of _____ cash and worry about paying the bill later.
                     (8)

A: I suppose you're right. Last month we bought _____ new flat
                                                       (9)

screen TV. We were at the store looking for a DVD player when we

saw it. It's so much nicer than our old TV, so we decided to get it and

put our _____ old TV in _____ basement. I suppose
            (10)   *(11)*

we didn't really need it.

B: Last weekend my husband bought _____ new CD
                                                    (12)

player. And he bought _____ new CDs. I asked him what
                              (13)

was wrong with our old CD player, and he said that it only played

two CDs at a time. _____ new CD player has room for 10 CDs.
                        (14)

A: Well, when we complain about our kids, we should realize that they are imitating us.

B: We need to make _____ changes in our own behavior. I'm
                        (15)

going to start _____ budget tonight. I'm going to start saving
                   (16)

_____ money each month.
     (17)

A: Me too.

14.6 | General or Specific with Quantity Words

If we put *of* the after a quantity word *(all, most, some,* etc.), we are making something specific. Without *of the*, the sentence is general.

Examples	Explanation
General: **All** children like toys. **Most** American homes have a television. **Many** teenagers have jobs. **Some** people are very rich.	We use *all, most, many, some, (a) few,* and *(a) little* before general nouns.

Very few people are billionaires.
 	
Specific:
 	
a. **All (of) the students** in this class have a textbook.
b. **Most of the students** in my art class have talent.
c. **Many of the topics** in this book are about life in America.
d. **Some of the people** in my building come from Haiti.
e. **Very few of the students** in this class are American citizens.
f. **Very little of the time** spent in this class is for reading.
g. **None of the classrooms** at this school has a telephone.
	We use *all of the, most of the, many of the, some of the, (a) few of the, (a) little of the,* and *none of the* before specific nouns.
	After *all, of* is often omitted.
	    All the students in this class
	       have a textbook.
	After *none of the* + plural noun, a singular verb is correct. However, you will often hear a plural verb used.
	    None of the classrooms ***have*** a
	       telephone.

Language Note:

Remember the difference between *a few* and *(very) few, a little* and *(very) little.* When we omit *a,* the emphasis is on the negative. We are saying the quantity is not enough. (See Lesson 5, Section 5.14 for more information.)

    Few people wanted to have a party. The party was canceled.

    A few people came to the meeting. We discussed our plans.

EXERCISE **10**   Fill in the blanks with *all, most, some,* or *(very) few* to make a general statement about Americans. Discuss your answers.

EXAMPLE   _____Most_____   Americans have a car.

1. _____ Americans have educational opportunities.
2. _____ Americans have a TV.
3. _____ American families have more than eight children.
4. _____ Americans know where my native country is.
5. _____ Americans shake hands when they meet.
6. _____ Americans use credit cards.
7. _____ Americans are natives of America.
8. _____ American citizens can vote.
9. _____ Americans speak my native language.
10. _____ Americans are unfriendly to me.

EXERCISE **11**   ABOUT YOU Fill in the blanks with a quantity word to make a true statement about specific nouns. If you use *none,* change the verb to the singular form.

EXAMPLES   _All of the_____ students in this class want to learn English.

s
None of the students in this class come∧from Australia.

1. _____ students in this class speak Spanish.
2. _____ students brought their books to class today.
3. _____ students are absent today.
4. _____ students want to learn English.
5. _____ students have jobs.
6. _____ students are married.
7. _____ students are going to return to their native countries.
8. _____ lessons in this book end with a review.
9. _____ pages in this book have pictures.
10. _____ tests in this class are hard.

Before You Read

1. Does American money look different from money in another country (size, color, etc.)?

2. Compare a one-dollar bill to a twenty-dollar bill. Do you see differences in design?

 Read the following article. Pay special attention to *other* and *another*.

The appearance of the American dollar did not change for a long time—from 1928 to 1996. But with advances in technology in recent years, it has become easier for counterfeiters[3] to copy dollar bills, making frequent changes necessary.

Did You Know?

Before 1928, the U.S. dollar was much bigger than the dollars we use today. The size of the dollar was reduced to save money on paper.

Look at the two twenty-dollar bills above. (Or see if you and your classmates have old and new bills.) You can see that on one twenty-dollar bill, the picture of Andrew Jackson is in an oval. On **the other** one, the picture is not in an oval. One bill has no background. **The other** bill has an eagle on the left and the words "Twenty USA" on the right. **Another** important change is in the color. The old bills are green. The new ones have some color. In the lower right corner of the old bill, the number "20" is in green. On the new bill, the "20" changes from gold to green, depending on how the light hits it. There are **other** changes too. If you have an old and a new bill, try to find **the other** differences.

The latest change to the U.S. bills began in 2003. The government decided to change the appearance of the twenty-dollar bill first, then the fifty- and one hundred-dollar bills. It has not been decided if the five- and ten-dollar bills will be changed. **The other** two bills ($1 and $2) will not be changed. Counterfeiters are not interested in small amounts of money. As new bills come into use, the old ones are "retired."

Some aspects of the bills remain the same: size, paper, the pictures on the front and back, and the motto "In God We Trust." In order to stay ahead of counterfeiters, the U.S. Treasury plans to redesign new bills every seven to ten years.

(continued)

---

[3] A *counterfeiter* is a person who makes copies of bills illegally.

U.S. Dollar Bills		
Denomination	**Front Side**	**Back Side**
$1	George Washington	Great Seal of the United States
$2	Thomas Jefferson	Declaration of Independence
$5	Abraham Lincoln	Lincoln Memorial
$10	Alexander Hamilton	Treasury Building
$20	Andrew Jackson	White House
$50	Ulysses S. Grant	U.S. Capitol
$100	Benjamin Franklin	Independence Hall

14.7 | *Another* and *Other*

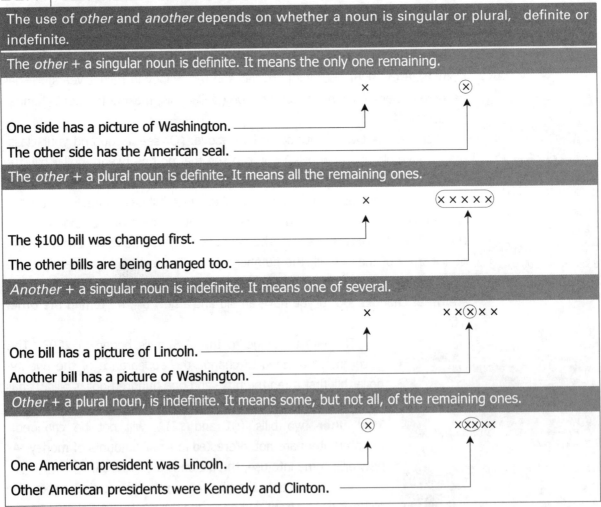

The use of *other* and *another* depends on whether a noun is singular or plural, definite or indefinite.

The *other* + a singular noun is definite. It means the only one remaining.

One side has a picture of Washington.
The other side has the American seal.

The *other* + a plural noun is definite. It means all the remaining ones.

The $100 bill was changed first.
The other bills are being changed too.

Another + a singular noun is indefinite. It means one of several.

One bill has a picture of Lincoln.
Another bill has a picture of Washington.

Other + a plural noun, is indefinite. It means some, but not all, of the remaining ones.

One American president was Lincoln.
Other American presidents were Kennedy and Clinton.

14.8 | More About *Another* and *Other*

Examples	Explanation
One change is the color. Another **one** is the frame around the face. The two-dollar bill is not common. The other **ones** are common.	We can use pronouns, *one* or *ones*, in place of the noun.
The two-dollar bill is not common. The **others** are common.	When the plural noun or pronoun (*ones*) is omitted, change *other* to *others*.
I have two bank accounts. One is for savings. **My other** account is for checking.	*The* is omitted when we use a possessive form. *Wrong:* My *the* other account is for checking.
I'm busy now. Can you come **another** time? Can you come **any other** time? Can you come **some other** time?	After *some* or *any, another* is changed to *other*. *Wrong:* Can you come *any* another time?
This dollar bill is old. You can't put it in the vending machine. You have to use **another** one.	*Another* is sometimes used to mean a different one.

EXERCISE 12   Fill in the blanks with *the other, another, the others, others,* or *other*.

EXAMPLE   One side of the one-dollar bill has a picture of George Washington.

_____The other_____ side has a picture of the American seal.

1. Some bills were changed in 2003. _____ bills were changed in 2004. Not all bills were changed.

2. Franklin, on the $100 bill, and Hamilton, on the $10 bill, were not American presidents. All _____ bills have pictures of American presidents.

3. Franklin was an important person in American history.

   _____ important people were Thomas Jefferson and John Hancock.

4. One bill has a picture of Lincoln. _____ one has a picture of George Washington.

5. George Washington was an American president. _____ presidents were Lincoln, Roosevelt, and Truman.

6. There were two presidents named Roosevelt. One was Theodore Roosevelt. _____ was Franklin Roosevelt.

7. The child has a lot of toys, but he wants _____ one.

8. New York is one big city in the U.S. _____ big cities are Philadelphia, Houston, and Detroit.

9. Many cities in the U.S. have warm weather. One city is Miami. _____ one is San Diego.

10. Do you know the capital of this state? Do you know _____ 49 state capitals?

Front

Card Inc.

1234 2123 8765 2334
Valid Dates
08/04–07/06
Jane A. Doe   credit

Jane A. Doe 5570 123

Back

11. Johnson is a common last name in the U.S. _____ common last names are Smith, Wilson, and Jones.

12. One side of the credit card has a name and number. _____ side has a place to sign your name.

13. If I use the ATM at my bank, I don't have to pay a fee. If I use it at any _____ bank, I have to pay a fee.

14. The bank is going to close now. Please come back some _____ time.

15. *Money* is a noncount noun. _____ ones are *love*, *freedom*, and *time*.

16. Some kids get an allowance for doing nothing. _____ kids have to do chores to get an allowance. But not all kids get an allowance.

17. We celebrate two presidents' birthdays in February. One is Lincoln's birthday. _____ is Washington's birthday,

18. The child gets presents from his grandparents. One grandparent is dead, but _____ three are alive.

EXERCISE **13**

A grandson (GS) and grandfather (GF) are talking. Fill in the blanks with *the other, another, the others,* or *others.*

GS: I want to buy ____another____ pair of sneakers.
             (example)

GF: What?! You already have about six pairs of sneakers. In fact, I bought you a new pair last month for your birthday.

GS: The new pair is fine, but _____ five are too small
                                              (1)
for me. You know I'm growing very fast, so I threw them away.

GF: Why did you throw them away? _____ boys in your
                                                        (2)
neighborhood could use them.

GS: They wouldn't like them. They're out of style.

GF: You kids are so wasteful today. What's wrong with the sneakers I bought you last month? If they fit you, why do you need _____ _____ pair?
(3)

GS: Everybody in my class at school has red sneakers with the laces tied backward. The sneakers you gave me are not in style any more.

GF: Do you always have to have what all _____ kids in
(4)

school have? Can't you think for yourself?

GS: Didn't you ask your parents for stuff when you were in junior high?

GF: My parents were poor, and my two brothers and I worked to help them. When we couldn't wear our clothes anymore because we outgrew them, we gave them to _____ families
(5)

nearby. And our neighbors gave us the things that their children outgrew. One neighbor had two sons. One son was a year older than me. _____ one was two years younger. So we
(6)

were constantly passing clothes back and forth.

GS: What about style? When clothes went out of style, didn't you throw them out?

GF: No. We never threw things out. Styles were not as important to us then. We didn't waste our parents' money thinking of styles. In fact, my oldest brother worked in a factory and gave all his salary to our parents. My _____ brother and I helped our father
(7)

in his business. My dad didn't give us a salary or an allowance. It was our duty to help him.

GS: You don't understand how important it is to look like all _____ kids.
(8)

GF: I guess I don't. I'm old fashioned. Every generation has _____ way of looking at things.
(9)

Before You Read

1. Have you received any financial aid to take this course?

2. Do you know how much it costs to get a college degree in the U.S.?

Read the following conversation between a son (S) and a dad (D). Pay attention to *one*, *some*, *any* (indefinite pronouns) and *it* and *them* (definite pronouns).

Did You Know?

In 2002, about 60 percent of undergraduates received some form of financial aid: grants, loans, and scholarships.

S: I decided not to go to college, Dad.

D: What? Do you know how important a college education is?

S: College is expensive. Besides, if I don't go to college now, I can start making money immediately. As soon as I earn **some**, I'd like to buy a car. Besides, my friends aren't going to college.

D: I'm not concerned about **them**. I'm interested in you and your future. I was just reading an article in a magazine about how much more money a college graduate earns than a high school graduate. Here's the article. Look at **it**. It says, "According to U.S. Census Bureau statistics, people with a bachelor's degree earn over 60 percent more than those with only a high school diploma. Over a lifetime, the gap in earning potential between a high school diploma and a B.A. (or higher) is more than $1,000,000."

S: Wow. I never realized that I could earn much more with a college degree than without **one**. Look. But the article also says, "In the 2003-2004 school year, the average tuition at a four-year private college was $27,000, and at a four-year public college, it was $10,000." How can you afford to send me to college?

D: I didn't just start to think about your college education today. I started to think about **it** when you were born. We saved money each month to buy a house, and we bought **one**. And we saved **some** each month for your college tuition.

S: That's great, Dad.

D: I also want you to apply for financial aid. There are grants, loans, and scholarships you should also look into. Your grades are good. I think you should apply for a scholarship.

S: I'll need to get an application.

D: I already thought of that. I brought **one** home today. Let's fill **it** out together.

S: Dad, if a college degree is so important to you, why didn't you get **one?**

D: When I was your age, we didn't live in the U.S. We were very poor and had to help our parents. You have a lot of opportunities for grants and scholarships, but we didn't have **any** when I was young.

S: Thanks for thinking about this from the day I was born.

Grants and Scholarships

Grants and scholarships provide aid that does not have to be repaid. However, some require that recipients maintain good grades or take certain courses.

Loans

Loans are another type of financial aid and are available to both students and parents. Like a car loan or a mortgage for a house, an education loan must eventually be repaid. Often, payments do not begin until the student finishes school. The interest rate on education loans is commonly lower than for other types of loans.

Source: http://www.ed.gov/pubs

Amount You Would Need to Save to Have $10,000 Available When Your Child Begins College (Assuming a 5 percent interest rate.)					
				Amount Available When Child Begins College	
If you start saving when your child is	Number of years of saving	Approximate monthly savings	Principal	Interest earned	Total savings
Newborn	18	$29	$6, 197	$3, 803	$10, 000
Age 4	14	41	6, 935	3, 065	10, 000
Age 8	10	64	7, 736	2, 264	10, 000
Age 12	6	119	8, 601	1, 399	10, 000
Age 16	2	397	9, 531	469	10, 000

Source of chart: http://www.ed.gov/pubs/Prepare/pt4.html

14.9 | Definite and Indefinite Pronouns

Examples	Explanation
I've always thought about <u>your education.</u> I started to think about **it** when you were born. I received <u>two college applications.</u> I have to fill **them** out. The father wants <u>his son</u> to go to college. The father is going to help **him.**	We use definite pronouns *him, her, them,* and *it* to refer to definite count nouns.
<u>A college degree</u> is important. It's hard to make a lot of money without **one.** I don't have <u>a scholarship.</u> I hope I can get **one.**	We use the indefinite pronoun *one* to refer to an indefinite singular count noun.
a. The father knew it was important to save <u>money.</u> He saved **some** every month. b. You received <u>five brochures</u> for colleges. Did you read **any?** c. You have a lot of <u>opportunities</u> today. When I was your age, we didn't have **any.**	We use *some* (for affirmative statements) and *any* (for negative statements and questions) to refer to an indefinite noncount noun (a) or an indefinite plural count noun (b) and (c).

Language Note:
We often use *any* and *some* before *more.*
> Dad, I don't have enough money. I need **some more.**
> Son, I'm not going to give you **any more.**

EXERCISE **14** A mother (M) is talking to her teenage daughter (D) about art school. Fill in the blanks with *one* or *it.*

M: I have some information about the state university. Do you want to look at _____*it*_____ with me?
(example)

D: I don't know, Mom. I don't know if I'm ready to go to college when I graduate.

M: Why not? We've been planning for _____ since the day you
(1)
were born.

D: College is not for everyone. I want to be an artist.

M: You can go to college and major in art. I checked out information about the art curriculum at the state university. It seems to have a very good program. Do you want to see information about _____?
(2)

D: I'm not really interested in college. To be an artist, I don't need a college degree.

M: But it's good to have _____ anyway.
　　　　　　　　　　　　　　　(3)

D: I don't know why. In college, I'll have to study general courses, too, like math and biology. You know I hate math. I'm not good at

_____ .
　　(4)

M: Well, maybe we should look at art schools. There's one downtown. Do you want to visit _____?
　　　　　　　　　　　　　　　　(5)

D: Yes, I'd like to. We can probably find information about _____ on the Web too.
　　(6)

(looking at the art school's Web site)

D: This school sounds great. Let's call and ask for an application.

M: I think you can get _____ online. Oh, yes, here it is.
　　　　　　　　　　　　(7)

D: Let's make a copy of _____ .
　　　　　　　　　　　　　　(8)

M: You can fill _____ out online and submit _____
　　　　　　　　(9)　　　　　　　　　　　　　　　(10)

electronically.

EXERCISE **15**   ABOUT YOU Answer each question.   Substitute the underlined words with an indefinite pronoun (*one, some, any*) or a definite pronoun (*it, them*).

EXAMPLES   Do you have a pen with you?
Yes, I have one.

Are you using your pen now?
No. I'm not using it now.

1. Does this school have a library?
2. How often do you use the library?
3. Do you have a dictionary?
4. When do you use your dictionary?
5. Did you buy any textbooks this semester?
6. How much did you pay for your textbooks?
7. Did the teacher give any homework last week?
8. Where did you do the homework?
9. Do you have any American neighbors?
10. Do you know your neighbors?
11. Does this college have a president?
12. Do you know the college president?
13. Did you receive any mail today?
14. What time does your letter carrier deliver your mail?

EXERCISE **16**   *Combination Exercise.*   This is a conversation between a teenage girl (A) and her mother (B). Fill in the blanks with *one, some, any, it, them, a, an, the,* or Ø (for no article).

A: Can I have 15 dollars?

B: What for?

A: I have' to buy _____a_____ poster of my favorite singer.
                        (example)

B: I gave you _____ money last week. What did you do with
                        (1)

_____?
        (2)

A: I spent _____ on a CD.
                (3)

B: No, you can't have _____ more money until next week.
                        (4)

A: Please, please, please. All of my friends have _____ . I'll
                                                        (5)

die if I don't get _____ .
                        (6)

B: What happened to all _____ money Grandpa gave you for
your birthday?   *(7)*

A: I spent _____ .

(8)

B: What about _____ money you put in the bank after your

(9)

graduation?

A: I don't have _____ more money in the bank.

(10)

B: You have to learn that _____ money doesn't grow on trees.

(11)

If you want me to give you _____ , you'll have to work for

(12)

it. You can start by cleaning your room.

A: But I cleaned _____ two weeks ago.

(13)

B: Two weeks ago was two weeks ago. It's dirty again.

A: I don't have _____ time. I have to meet my friends.

(14)

B: You can't go out. You need to do your homework.

A: I don't have _____ . Please let me have 15 dollars.

(15)

B: When I was your age, I had _____ job. And I gave my

(16)

parents half of _____ money I earned. You kids today

(17)

have _____ easy life.

(18)

A: Why do _____ parents always say that to _____

(19)   (20)

kids?

B: Because it's true. It's time you learn that _____ life is hard.

(21)

A: I bet Grandpa said that to you when you were _____ child.

(22)

B: And I bet you'll say it to your kids when you're _____ adult.

(23)

1. Articles

	Count—Singular	Count—Plural	Noncount
General	*A/An* **A child** likes toys.	Ø Article **Children** like toys. I love **children.**	Ø Article **Money** can't buy happiness. Everyone needs **money.**
Indefinite	*A/An* I bought **a toy.**	*Some/Any* I bought **some toys.** I didn't buy **any games.**	*Some/Any* I spent **some money.** I didn't buy **any candy.**
Specific	*The* **The toy** on the floor is for the baby.	*The* **The toys** on the table are for you.	*The* **The money** on the table is mine.
Classification	*A/An* A toddler is **a young child.**	Ø Article Teenagers are **young adults.**	————

2. *Other / Another*

	Definite	Indefinite
Singular	□□ ⊂□⊃ the other book my other book the other one the other	□□ □□⊂□⊃□□ □□ □□ another book some/any other book another one another
Plural	□□⊂□ □□ □□ □□ □□⊃ the other books my other books the other ones the others	□□ □□ ⊂□□ □□ □□⊃□□ other books some/any other books other ones others

3. Indefinite Pronouns—Use *one / some / any* to substitute for indefinite nouns.

I need a quarter. Do you have **one?**

I need some pennies. You have **some.**

I don't have any change. Do you have **any?**

EDITING ADVICE

1. Use *the* after a quantity word when the noun is definite.

   the
   Most of ∧students in my class are from Romania.

2. Be careful with *most* and *almost*.

   Most of
   ~~Almost~~ my teachers are very patient.

3. Use a plural count noun after a quantity expression.

   s
   A few of my friend ∧live in Canada.

4. *Another* is always singular.

   Other
   Some teachers are strict. ~~Another~~ teachers are easy:

5. Use an indefinite pronoun to substitute for an indefinite noun.

   one
   I need to borrow a pen. I didn't bring ~~it~~ today.

6. A and *an* are always singular.

   She has ~~a~~ beautiful eyes.

7. Don't use *there* to introduce a unique, definite noun.

   T   is
   ~~There's~~ the Statue of Liberty ∧in New York.

8. Use *a* or *an* for a definition or a classification of a singular count noun.

   a
   The Statue of Liberty is ∧monument.

9. Don't use *the* with a possessive form.

   One of my sisters lives in New York. My ~~the~~ other sister lives in New Jersey.

PART **1** Find the mistakes with the underlined words, and correct them. Not every sentence has a mistake. If the sentence is correct, write *C*.

EXAMPLES One of her classmates is from Mexico. ~~Other~~ *Another* one is from Spain.

Most Americans own a TV. C

1. All of teachers at this college have a master's degree.

2. Some of the animals eat only meat. They are called "carnivores."

3. The students in this class come from many countries. Some of the students are from Poland. Another students are from Hungary.

4. A battery has two terminals. One is positive; another is negative.

5. I'm taking two classes. One is English. The other is math.

6. I lost my dictionary. I need to buy another one.

7. I lost my textbook. I think I lost it in the library.

8. I don't have a computer. Do you have it?

9. Most my teachers have a lot of experience.

10. Cuba is country.

11. Most women want to have children.

12. I have some money with me. Do you have any?

13. Very few of the students in this class have financial aid. Most of us pay tuition.

14. I have two brothers. One of my brothers is an engineer. The other my brother is a physical therapist.

15. Almost my friends come from South America.

16. There's the Golden Gate Bridge in San Francisco.

PART **2** Fill in the blanks with *the, a, an, some, any,* or *Ø* (for no article). In some cases, more than one answer is possible.

A: Do you want to come to my house tonight? I rented ____*some*____
(*example*)
movies. We can make _____ popcorn and watch
(1)
_____ movies together.
(2)

B: Thanks, but I'm going to _____ party. Do you want to
                                    (3)

go with me?

A: Where's it going to be?

B: It's going to be at Michael's apartment.

A: Who's going to be at _____ party?
                                    (4)

B: Most of _____ students in my English class will be
                (5)

there. Each student is going to bring _____ food.
                                                (6)

A: _____ life in the U.S. is strange. In my country,
          (7)

_____ people don't have to bring _____
          (8)   *(9)*

food to a party.

B: That's the way it is in my country, too. But we're in _____
                                                                    (10)

U.S. now. I'm going to bake _____ cake. You can make
                                      (11)

_____ special dish from your country.
          (12)

A: You know I'm _____ terrible cook.
                          (13)

B: Don't worry. You can buy something. My friend Max is going to buy

_____ crackers and cheese. Why don't you bring
          (14)

_____ salami or roast beef?
          (15)

A: But I don't eat _____ meat. I'm _____
                          (16)   *(17)*

vegetarian.

B: Well, you can bring _____ bowl of fruit.
                                (18)

A: That's _____ good idea. What time does _____
                (19)   *(20)*

party start?

B: At 8 o'clock.

A: I have to take my brother to _____ airport at 6:30. I
                                          (21)

don't know if I'll be back on time.

B: You don't have to arrive at 8 o'clock exactly. I'll give you

_____ address, and you can arrive any time you want.
          (22)

PART **3** Fill in the blanks with *other, others, another, the other,* or *the others.*

A: I don't like my apartment.

B: Why not?

A: It's very small. It only has two closets. One is big, but
_____the other_____ is very small.
     (example)

B: That's not very serious. Is that the only problem? Are there _____
_____ problems?
       (1)

A: There are many _____.
             (2)

B: Such as?

A: Well, the landlord doesn't provide enough heat in the winter.

B: Hmm. That's a real problem. Did you complain to him?

A: I did, but he says that all _____ tenants are happy.
                (3)

B: Why don't you look for _____ apartment?
              (4)

A: I have two roommates. One wants to move, but _____
                                         (5)

likes it here.

B: Well, if one wants to stay and _____ two want to move,
                         (6)

why don't you move and look for_____ roommate?
                           (7)

PART **4** Fill in the blanks with *one, some, any, it,* or *them.*

EXAMPLE I have a computer, but my roommate doesn't
have _____one_____.

1. Do you want to use my bicycle? I won't need _____ this afternoon.

2. I rented a movie. We can watch _____ tonight.

3. My English teacher gives some homework every day, but she doesn't give _____ on the weekends.

4. My class has a lot of Mexican students. Does your class have _____
_____?

5. I wrote two compositions last week, but I got bad grades because I wrote _____ very quickly.

6. I don't have any problems with English, but my roommate has _____ _____.

7. I can't remember the teacher's name. Do you remember _____?

8. You won't need any paper for the test, but you'll need _____ for the composition.

9. I went to the library to find some books in my language, but I couldn't find _____.

EXPANSION ACTIVITIES

Classroom *Activities*

1. Fill in the blanks with *all, most, some, a few,* or *very few* to make a general statement about your native country or another country you know well. Find a partner from a different country, if possible, and compare your answers.

a. _____ banks are safe places to put your money.
b. _____ doctors make a lot of money.
c. _____ teenagers work.
d. _____ children work.
e. _____ teachers are rich.
f. _____ government officials are rich.
g. _____ children get an allowance.
h. _____ people work on Saturdays.
i. _____ businesses are closed on Sundays.
j. _____ families own a car.
k. _____ women work outside the home.
l. _____ people have a college education.
m. _____ people have servants.
n. _____ married couples have their own apartment.
o. _____ old people live with their grown children.
p. _____ people speak English.
q. _____ children study English in school.
r. _____ parents have more than five children.

s. _____ people live in an apartment.

t. _____ young men serve in the military.

u. _____ people are happy with the political situation.

2. Bring in coins and bills from your native country or another country you've visited. Form a small group of students from different countries, and show this money to the other students in your group.

Talk About it

1. The following sayings and proverbs are about money. Discuss the meaning of each one. Do you have a similar saying in your native language?
   - All that glitters isn't gold.
   - Money is the root of all evil.
   - Friendship and money don't mix.
   - Another day, another dollar.
   - Money talks.

2. Discuss ways to save money. Discuss difficulties in saving money.

3. **Discuss this saying:** The difference between men and boys is the price of their toys.

Write About it

Do you think kids should get an allowance from their parents? How much? Does it depend on the child's age? Should the child have to work for the money? Write a few paragraphs.

Internet Activities

1. Look for bank rates on the Internet. Compare the interest on a one-year CD (certificate of deposit) at two banks.

2. Find a currency converter on the Web. Convert the American dollar to the currency of another country.

3. Go online to find an application for financial aid. Do you have any questions on how to fill it out?

4. Find the Web site of a college or university you are interested in. Find out the cost of tuition.

Additional Activities at http://elt.thomson.com/gic

Appendices

APPENDIX A

Spelling and Pronunciation of Verbs

Spelling of the -s Form of Verbs		
Rule	Base Form	-s Form
Add s to most verbs to make the -s form.	hope eat	hopes eats
When the base form ends in s, z, sh, ch, or x, add es and pronounce an extra syllable, /əz/.	miss buzz wash catch fix	misses buzzes washes catches fixes
When the base form ends in a consonant + y, change the y to i and add es.	carry worry	carries worries
When the base form ends in a vowel + y, do not change the y.	pay obey	pays obeys
Add es to go and do.	go do	goes does

Three Pronunciations of the -s Form		
We pronounce /s/ if the verb ends in these voiceless sounds: /p t k f/.	hope—hopes eat—eats	pick—picks laugh—laughs
We pronounce /z/ if the verb ends in most voiced sounds.	live—lives grab—grabs read—reads	run—runs sing—sings borrow—borrows
When the base form ends in s, z, sh, ch, x, se, ge, or ce, we pronounce an extra syllable, /əz/.	miss—misses buzz—buzzes wash—washes	fix—fixes use—uses change—changes
These verbs have a change in the vowel sound.	watch—watches do/du/—does/dəz/	dance—dances say/**sei**/—says/s**e**z/

Spelling of the *-ing* Form of Verbs		
Rule	Base Form	*-ing* Form
Add *-ing* to most verbs. Note: Do not remove the *y* for the *-ing* form.	eat go study	eating going studying
For a one-syllable verb that ends in a consonant + vowel + consonant (CVC), double the final consonant and add *-ing*.	p l a n | | | C V C s t o p | | | C V C s i t | | | C V C	planning stopping sitting
Do not double the final *w, x,* or *y*.	show mix stay	showing mixing staying
For a two-syllable word that ends in CVC, double the final consonant only if the last syllable is stressed.	refér admít begín	referring admitting beginning
When the last syllable of a two-syllable word is not stressed, do not double the final consonant.	lísten ópen óffer	listening opening offering
If the word ends in a consonant + *e*, drop the *e* before adding *-ing*.	live take write	living taking writing

Spelling of the Past Tense of Regular Verbs

Rule	Base Form	-ed Form
Add *ed* to the base form to make the past tense of most regular verbs.	start kick	started kicked
When the base form ends in *e*, add *d* only.	die live	died lived
When the base form ends in a consonant + *y*, change the *y* to *i* and add *ed*.	carry worry	carried worried
When the base form ends in a vowel + *y*, do not change the *y*.	destroy stay	destroyed stayed
For a one-syllable word that ends in a consonant + vowel + consonant (CVC), double the final consonant and add *ed*.	s t o p | | | C V C p l u g | | | C V C	stopped plugged
Do not double the final *w* or *x*.	sew fix	sewed fixed
For a two-syllable word that ends in CVC, double the final consonant only if the last syllable is stressed.	occúr permít	occurred permitted
When the last syllable of a two-syllable word is not stressed, do not double the final consonant.	ópen háppen	opened happened

Pronunciation of Past Forms that End in -ed

The past tense with -*ed* has three pronunciations.

We pronounce a /**t**/ if the base form ends in these voiceless. sounds: /**p**, **k**, **f**, **s**, š, č/.	jump—jumped cook—cooked	cough—coughed kiss—kissed	wash—washed watch—watched
We pronounce a /**d**/ if the base form ends in most voiced sounds.	rub—rubbed drag—dragged love—loved bathe—bathed use—used	charge—charged glue—glued massage—massaged name—named learn—learned	bang—banged call—called fear—feared free—freed
We pronounce an extra syllable /ə**d**/ if the base form ends in a /**t**/ or /**d**/ sound.	wait—waited hate—hated	want—wanted add—added	need—needed decide—decided

Irregular Noun Plurals

Singular	Plural	Explanation
man woman mouse tooth foot goose	men women mice teeth feet geese	Vowel change (Note: The first vowel in *women* is pronounced /ɪ/.)
sheep fish deer	sheep fish deer	No change
child person	children people (OR persons)	Different word form
	(eye) glasses belongings clothes goods groceries jeans pajamas pants/slacks scissors shorts	No singular form
alumnus cactus radius stimulus syllabus	alumni cacti OR cactuses radii stimuli syllabi OR syllabuses	*us* → *i*
analysis crisis hypothesis oasis parenthesis thesis	analyses crises hypotheses oases parentheses theses	*is* → *es*

Continued

Singular	Plural	Explanation
appendix index	appendices OR appendixes indices OR indexes	$ix \rightarrow ices$ OR $\rightarrow ixes$
bacterium curriculum datum medium memorandum criterion phenomenon	bacteria curricula data media memoranda criteria phenomena	$um \rightarrow a$ $ion \rightarrow a$ $on \rightarrow a$
alga formula vertebra	algae formulae OR formulas vertebrae	$a \rightarrow ae$

APPENDIX C

Spelling Rules for Adverbs Ending in *-ly*

Adjective Ending	Examples	Adverb Ending	Adverb
Most endings	careful quiet serious	Add -*ly*.	carefully quietly seriously
-*y*	easy happy lucky	Change *y* to *i* and add -*ly*.	easily happily luckily
-*e*	nice free	Keep the *e* and add -*ly*.*	nicely freely
consonant + *le*	simple comfortable double	Drop the *e* and add -*ly*.	simply comfortably doubly
-*ic*	basic enthusiastic	Add -*ally*.**	basically enthusiastically

Exceptions:
  * true—truly
  ** public—publicly			

Metric Conversion Chart

LENGTH

When You Know	Symbol	Multiply by	To Find	Symbol
inches	in	2.54	centimeters	cm
feet	ft	30.5	centimeters	cm
feet	ft	0.3	meters	m
yards	yd	0.91	meters	m
miles	mi	1.6	kilometers	km
Metric:				
centimeters	cm	0.39	inches	in
centimeters	cm	0.03	feet	ft
meters	m	3.28	feet	ft
meters	m	1.09	yards	yd
kilometers	km	0.62	miles	mi

Note:

1 foot = 12 inches; 1 yard = 3 feet or 36 inches

AREA

When You Know	Symbol	Multiply by	To Find	Symbol
square inches	in^2	6.5	square centimeters	cm^2
square feet	ft^2	0.09	square meters	m^2
square yards	yd^2	0.8	square meters	m^2
square miles	mi^2	2.6	square kilometers	km^2
Metric:				
square centimeters	cm^2	0.16	square inches	in^2
square meters	m^2	10.76	square feet	ft^2
square meters	m^2	1.2	square yards	yd^2
square kilometers	km^2	0.39	square miles	mi^2

WEIGHT (Mass)				
When You Know	Symbol	Multiply by	To Find	Symbol
ounces	oz	28.35	grams	g
pounds	lb	0.45	kilograms	kg
Metric:				
grams	g	0.04	ounces	oz
kilograms	kg	2.2	pounds	lb

Note:
16 ounces =1 pound

VOLUME				
When You Know	Symbol	Multiply by	To Find	Symbol
fluid ounces	fl oz	30.0	milliliters	ml
pints	pt	0.47	liters	l
quarts	qt	0.95	liters	l
gallons	gal	3.8	liters	l
Metric:				
milliliters	ml	0.03	fluid ounces	fl oz
liters	l	2.11	pints	pt
liters	l	1.05	quarts	qt
liters	l	0.26	gallons	gal

TEMPERATURE				
When You Know	Symbol	Do This	To Find	Symbol
degrees Fahrenheit	°F	Subtract 32, then multiply by 5/9	degrees Celsius	°C
Metric:				
degrees Celsius	°C	Multiply by 9/5, then add 32	degrees Fahrenheit	°F

Sample temperatures:			
Fahrenheit	Celsius	Fahrenheit	Celsius
0	–18	60	16
10	–12	70	21
20	–7	80	27
30	–1	90	32
40	4	100	38
50	10	212	100

The Verb *Get*

Get has many meanings. Here is a list of the most common ones:

- get something = receive
  I got a letter from my father.

- get + (to) place = arrive
  I got home at six. What time do you get to school?

- get + object + infinitive = persuade
  She got him to wash the dishes.

- get + past participle = become

get acquainted	get worried	get hurt
get engaged	get lost	get bored
get married	get accustomed to	get confused
get divorced	get used to	get scared
get tired	get dressed	

  They got married in 1989.

- get + adjective = become

get hungry	get upset	get dark
get rich	get sleepy	get angry
get nervous	get fat	get old
get well		

  It gets dark at 6:30.

- get an illness = catch
  While she was traveling, she got malaria.

- get a joke or an idea = understand
  Everybody except Tom laughed at the joke. He didn't get it.
  The boss explained the project to us, but I didn't get it.

- get ahead = advance
  He works very hard because he wants to get ahead in his job.

- get along (well) (with someone) = have a good relationship
  She doesn't get along with her mother-in-law.
  Do you and your roommate get along well?

- get around to something = find the time to do something
  I wanted to write my brother a letter yesterday, but I didn't get around to it.

- get away = escape
  The police chased the thief, but he got away.

- get away with something = escape punishment

  He cheated on his taxes and got away with it.

- get back = return

  He got back from his vacation last Saturday.

- get back at someone = get revenge

  My brother wants to get back at me for stealing his girlfriend.

- get back to someone = communicate with someone at a later time

  The boss can't talk to you today. Can she get back to you tomorrow?

- get by = have just enough but nothing more

  On her salary, she's just getting by. She can't afford a car or a vacation.

- get in trouble = be caught and punished for doing something wrong

  They got in trouble for cheating on the test.

- get in(to) = enter a car

  She got in the car and drove away quickly.

- get out (of) = leave a car

  When the taxi arrived at the theater, everyone got out.

- get on = seat oneself on a bicycle, motorcycle, horse

  She got on the motorcycle and left.

- get on = enter a train, bus, airplane

  She got on the bus and took a seat in the back.

- get off = leave a bicycle, motorcycle, horse, train, bus, airplane

  They will get off the train at the next stop.

- get out of something = escape responsibility

  My boss wants me to help him on Saturday, but I'm going to try to get out of it.

- get over something = recover from an illness or disappointment

  She has the flu this week. I hope she gets over it soon.

- get rid of someone or something = free oneself of someone or something undesirable

  My apartment has roaches, and I can't get rid of them.

- get through (to someone) = communicate, often by telephone

  She tried to explain the harm of eating fast food to her son, but she couldn't get through to him.

  I tried to call my mother many times, but her line was busy. I couldn't get through.

- get through with something = finish

  I can meet you after I get through with my homework.

- get together = meet with another person

  I'd like to see you again. When can we get together?

- get up = arise from bed

  He woke up at 6 o'clock, but he didn't get up until 6:30.

Make and *Do*

Some expressions use *make*. Others use *do*.	
Make	Do
make a date/an appointment	do (the) homework
make a plan	do an exercise
make a decision	do the dishes
make a telephone call	do the cleaning, laundry, ironing, washing, etc.
make a reservation	do the shopping
make a mistake	do one's best
make an effort	do a favor
make an improvement	do the right/wrong thing
make a promise	do a job
make money	do business
make noise	What do you do for a living? (asks about a job)
make the bed	How do you do? (said when you meet someone for the first time)

Nouns That Can Be Both Count or Noncount

In the following cases, the same word can be a count or a noncount noun. The meaning is different, however.

Noncount	Count
I spent a lot of *time* on my project.	I go shopping two *times* a month.
I have a lot of *experience* with computers.	I had a lot of interesting *experiences* on my trip to Europe.

In the following cases, there is a small difference in meaning. We see a noncount noun as a whole unit. We see a count noun as something that can be divided into parts.

Noncount	Count
There is a lot of *crime* in a big city.	A lot of *crimes* are never solved.
There is a lot of *opportunity* to make money in the U.S.	There are a lot of job *opportunities* in my field.
She bought a lot of *fruit*.	Oranges and lemons are *fruits* that have a lot of Vitamin C.
I don't have much *food* in my refrigerator.	Milk and butter are *foods* that contain cholesterol.
I have a lot of *trouble* with my car.	He has many *troubles* in his life.

Verbs and Adjectives Followed by a Preposition

Many verbs and adjectives are followed by a preposition.

accuse someone of	(be) famous for	prevent (someone) from
(be) accustomed to	feel like	prohibit (someone) from
adjust to	(be) fond of	protect (someone) from
(be) afraid of	forget about	(be) proud of
agree with	forgive someone for	recover from
(be) amazed at/by	(be) glad about	(be) related to
(be) angry about	(be) good at	rely on/upon
(be) angry at/with	(be) grateful to someone for	(be) responsible for
apologize for	(be) guilty of	(be) sad about
approve of	(be) happy about	(be) satisfied with
argue about	hear about	(be) scared of
argue with	hear of	(be) sick of
(be) ashamed of	hope for	(be) sorry about
(be) aware of	(be) incapable of	(be) sorry for
believe in	insist on/upon	speak about
blame someone for	(be) interested in	speak to/with
(be) bored with/by	(be) involved in	succeed in
(be) capable of	(be) jealous of	(be) sure of/about
care about/for	(be) known for	(be) surprised at
compare to/with	(be) lazy about	take care of
complain about	listen to	talk about
(be) concerned about	look at	talk to/with
concentrate on	look for	thank (someone) for
consist of	look forward to	(be) thankful (to someone) for
count on	(be) mad about	think about/of
deal with	(be) mad at	(be) tired of
decide on	(be) made from/of	(be) upset about
depend on/upon	(be) married to	(be) upset with
(be) different from	object to	(be) used to
disapprove of	(be) opposed to	wait for
(be) divorced from	participate in	warn (someone) about
dream about/of	plan on	(be) worried about
(be) engaged to	pray to	worry about
(be) excited about	pray for	
(be) familiar with	(be) prepared for/to	

Direct and Indirect Objects

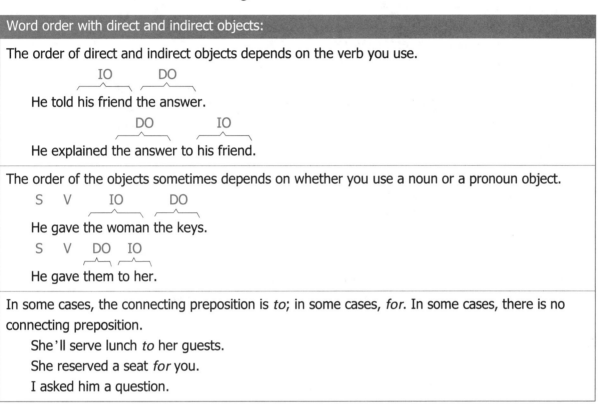

Word order with direct and indirect objects:

The order of direct and indirect objects depends on the verb you use.

        IO   DO

He told his friend the answer.

           DO   IO

He explained the answer to his friend.

The order of the objects sometimes depends on whether you use a noun or a pronoun object.

  S   V   IO   DO

He gave the woman the keys.

  S   V   DO IO

He gave them to her.

In some cases, the connecting preposition is *to*; in some cases, *for*. In some cases, there is no connecting preposition.

    She'll serve lunch *to* her guests.

    She reserved a seat *for* you.

    I asked him a question.

The order of direct and indirect objects depends on the verb you use. It also can depend on whether you use a noun or a pronoun as the object.

Group 1 Pronouns affect word order. The preposition used is *to*.

Patterns: He gave a present to his wife. (DO to IO)
He gave his wife a present. (IO/DO)
He gave it to his wife. (DO to IO)
He gave her a present. (IO/DO)
He gave it to her. (DO to IO)

Verbs:

bring	lend	pass	sell	show	teach
give	offer	pay	send	sing	tell
hand	owe	read	serve	take	write

Group 2 Pronouns affect word order. The preposition used is *for*.

Patterns: He bought a car for his daughter. (DO for IO)
He bought his daughter a car. (IO/DO)
He bought it for his daughter. (DO for IO)
He bought her a car. (IO/DO)
He bought it for her. (DO for IO)

Verbs:

bake	buy	draw	get	make
build	do	find	knit	reserve

Group 3 Pronouns don't affect word order. The preposition used is *to*.

Patterns: He explained the problem to his friend. (DO to IO)
He explained it to her. (DO to IO)

Verbs:

admit	introduce	recommend	say
announce	mention	repeat	speak
describe	prove	report	suggest
explain			

Group 4 Pronouns don't affect word order. The preposition used is *for*.

Patterns: He cashed a check for his friend. (DO for IO)
He cashed it for her. (DO for IO)

Verbs:

answer	change	design	open	prescribe
cash	close	fix	prepare	pronounce

Group 5 Pronouns don't affect word order. No preposition is used.

Patterns: She asked the teacher a question. (IO/DO)
She asked him a question. (IO/DO)
It took me five minutes to answer the question. (IO/DO)

Verbs:

ask	charge	cost	wish	take (with time)

Capitalization Rules

- The first word in a sentence: **My** friends are helpful.
- The word "I": My sister and **I** took a trip together.
- Names of people: **Michael Jordan; George Washington**
- Titles preceding names of people: **Doctor (Dr.) Smith; President Lincoln; Queen Elizabeth; Mr. Rogers; Mrs. Carter**
- Geographic names: the **United States; Lake Superior; California;** the **Rocky Mountains;** the **Mississippi River**
  NOTE: The word "the" in a geographic name is not capitalized.
- Street names: **Pennsylvania Avenue (Ave.); Wall Street (St.); Abbey Road (Rd.)**
- Names of organizations, companies, colleges, buildings, stores, hotels: the **Republican Party; Thomson Heinle; Dartmouth College;** the **University of Wisconsin;** the **White House; Bloomingdale's;** the **Hilton Hotel**
- Nationalities and ethnic groups: **Mexicans; Canadians; Spaniards; Americans; Jews; Kurds; Eskimos**
- Languages: **English; Spanish; Polish; Vietnamese; Russian**
- Months: **January; February**
- Days: **Sunday; Monday**
- Holidays: **Christmas; Independence Day**
- Important words in a title: **Grammar in Context; The Old Man and the Sea; Romeo and Juliet; The Sound of Music**
  NOTE: Capitalize "the" as the first word of a title.

Glossary of Grammatical Terms

- **Adjective** An adjective gives a description of a noun.
  It's a *tall* tree.   He's an *old* man.   My neighbors are *nice*.
- **Adverb** An adverb describes the action of a sentence or an adjective or another adverb.
  She speaks English *fluently*.   I drive *carefully*.
  She speaks English *extremely* well.   She is *very* intelligent.
- **Adverb of Frequency** An adverb of frequency tells how often the action happens.
  I *never* drink coffee.   They *usually* take the bus.

- **Affirmative** means *yes*.

- **Apostrophe**' We use the apostrophe for possession and contractions.
    My *sister's* friend is beautiful.   Today *isn't* Sunday.

- **Article** The definite article is *the*. The indefinite articles are *a* and *an*.
    I have *a* cat.   I ate *an* apple.   *The* president was late.

- **Auxiliary Verb** Some verbs have two parts: an auxiliary verb and a main verb.
    He *can't* study.   We *will* return.

- **Base Form** The base form, sometimes called the "simple" form of the verb, has no tense. It has no ending (*-s* or *-ed*): *be, go, eat, take, write*.
    He doesn't *know* the answer.   I didn't *go* out.
    You shouldn't *talk* loudly.

- **Capital Letter** A B C D E F G...

- **Clause** A clause is a group of words that has a subject and a verb.
    Some sentences have only one clause.
        She found a good job.

    Some sentences have a **main clause** and a **dependent clause**.

MAIN CLAUSE	DEPENDENT CLAUSE (**reason clause**)
She found a good job	because she has computer skills.
MAIN CLAUSE	DEPENDENT CLAUSE (**time clause**)
She'll turn off the light	before she goes to bed.
MAIN CLAUSE	DEPENDENT CLAUSE (*if* **clause**)
I'll take you to the doctor	if you don't have your car on Saturday.

- **Colon :**

- **Comma ,**

- **Comparative Form** A comparative form of an adjective or adverb is used to compare two things.
    My house is *bigger* than your house.
    Her husband drives *faster* than she does.

- **Complement** The complement of the sentence is the information after the verb. It completes the verb phrase.
    He works *hard*.   I slept *for five hours*.   They are *late*.

- **Consonant** The following letters are consonants: *b, c, d, f, g, h, j, k, l, m, n, p, q, r, s, t, v, w, x, y, z*.
    NOTE: *y* is sometimes considered a vowel, as in the word *syllable*.

- **Contraction** A contraction is made up of two words put together with an apostrophe.
    He's my brother.   *You're* late.   They *won't* talk to me.
    (*He's* = *He is*)   (*You're* = *You are*)   (*won't* = *will not*)

- **Count Noun** Count nouns are nouns that we can count. They have a singular and a plural form.

  1 pen — 3 pens   1 table — 4 tables

- **Dependent Clause** See **Clause**.

- **Direct Object** A direct object is a noun (phrase) or pronoun that receives the action of the verb.

  We saw *the movie*.   You have *a nice car*.   I love *you*.

- **Exclamation Mark** !

- **Frequency Words** Frequency words are *always, usually, often, sometimes, rarely, seldom,* and *never*.

  I *never* drink coffee.   We *always* do our homework.

- **Hyphen** –

- **Imperative** An imperative sentence gives a command or instructions. An imperative sentence omits the word *you*.

  Come here.   *Don't be* late.   Please *sit* down.

- **Indefinite Pronoun** An indefinite pronoun (*one, some, any*) takes the place of an indefinite noun.

  I have a cell phone. Do you have *one*?

  I didn't drink any coffee, but you drank *some*. Did he drink *any*?

- **Infinitive** An infinitive is *to* + base form.

  I want *to leave*.   You need *to be* here on time.

- **Linking Verb** A linking verb is a verb that links the subject to the noun or adjective after it. Linking verbs include *be, seem, feel, smell, sound, look, appear, taste*.

  She *is* a doctor.   She *seems* very intelligent.   She *looks* tired.

- **Modal** The modal verbs are *can, could, shall, should, will, would, may, might,* and *must*.

  They *should* leave.   I *must* go.

- **Negative** means *no*.

- **Nonaction Verb** A nonaction verb has no action. We do not use a continuous tense (*be* + verb *-ing*) with a nonaction verb. The nonaction verbs are: *believe, cost, care, have, hear, know, like, love, matter, mean, need, own, prefer, remember, see, seem, think, understand,* and *want*.

  She *has* a laptop.   We *love* our mother.

- **Noncount Noun** A noncount noun is a noun that we don't count. It has no plural form.

  She drank some *water*.   He prepared some *rice*.

  Do you need any *money*?

- **Noun** A noun is a person (*brother*), a place (*kitchen*), or a thing (*table*). Nouns can be either count (*1 table, 2 tables*) or noncount (*money, water*).

  My *brother* lives in California.   My *sisters* live in New York.

  I get *mail* from my family.

- **Noun Modifier** A noun modifier makes a noun more specific.

  fire department   *Independence* Day   *can* opener

- **Noun Phrase** A noun phrase is a group of words that form the subject or object of the sentence.

  A very nice woman helped me at registration.

  I bought *a big box of candy.*

- **Object** The object of the sentence follows a verb. It receives the action of the verb.

  He bought *a car.*   I saw *a movie.*   I met *your brother.*

- **Object Pronoun** Use object pronouns (*me, you, him, her, it, us, them*) after the verb or preposition.

  He likes *her.*   I saw the movie. Let's talk about *it.*

- **Paragraph** A paragraph is a group of sentences about one topic.

- **Parentheses ( )**

- **Participle, Present** The present participle is verb + *-ing*.

  She is *sleeping.*   They were *laughing.*

- **Period .**

- **Phrase** A group of words that go together.

  Last month my sister came to visit.

  There is a strange car *in front of my house.*

- **Plural** Plural means more than one. A plural noun usually ends with *-s.*

  She has beautiful *eyes.*

- **Possessive Form** Possessive forms show ownership or relationship.

  Mary's coat is in the closet.   *My* brother lives in Miami.

- **Preposition** A preposition is a short connecting word: *about, above, across, after, around, as, at, away, back, before, behind, below, by, down, for, from, in, into, like, of, off, on, out, over, to, under, up, with.*

  The book is *on* the table.

- **Pronoun** A pronoun takes the place of a noun:

  I have a new car. I bought *it* last week.

  John likes Mary, but *she* doesn't like *him.*

- **Punctuation** Period . Comma , Colon : Semicolon ; Question Mark ? Exclamation Mark !

- **Question Mark ?**

- **Quotation Marks " "**

- **Regular Verb** A regular verb forms its past tense with *-ed*.

    He *worked* yesterday.   I *laughed* at the joke.

- ***s* Form** A present tense verb that ends in *-s* or *-es*.

    He *lives* in New York.   She *watches* TV a lot.

- **Sense-Perception Verb** A sense-perception verb has no action.   It describes a sense.

    She *feels* fine.   The coffee *smells* fresh.   The milk *tastes* sour.

- **Sentence** A sentence is a group of words that contains a subject[1] and a verb (at least) and gives a complete thought.

    SENTENCE: She came home.

    NOT A SENTENCE: When she came home

- **Simple Form of Verb** The simple form of the verb, also called the base form, has no tense; it never has an *-s*, *-ed*, or *-ing* ending.

    Did you *see* the movie?   I couldn't *find* your phone number.

- **Singular** Singular means one.

    She ate a *sandwich*.   I have one *television*.

- **Subject** The subject of the sentence tells who or what the sentence is about.

    My sister got married last April. *The wedding* was beautiful.

- **Subject Pronouns** Use subject pronouns (*I, you, he, she, it, we, you, they*) before a verb.

    They speak Japanese.   *We* speak Spanish.

- **Superlative Form** A superlative form of an adjective or adverb shows the number one item in a group of three or more.

    January is the *coldest* month of the year.

    My brother speaks English the *best* in my family.

- **Syllable** A syllable is a part of a word that has only one vowel sound. (Some words have only one syllable.)

    change (one syllable)   after (af·ter = two syllables)

    look (one syllable)   responsible (re·spon·si·ble = four syllables)

- **Tag Question** A tag question is a short question at the end of a sentence. It is used in conversation.

    You speak Spanish, *don't you*?   He's not happy, *is he*?

- **Tense** A verb has tense.   Tense shows when the action of the sentence happened.

    SIMPLE PRESENT: She usually *works* hard.

    FUTURE: She *will work* tomorrow.

    PRESENT CONTINUOUS: She *is working* now.

    SIMPLE PAST: She *worked* yesterday.

---

[1] In an imperative sentence, the subject *you* is omitted: *Sit down. Come here.*

- **Verb** A verb is the action of the sentence.

    He *runs* fast.   I *speak* English.

    Some verbs have no action. They are linking verbs. They connect the subject to the rest of the sentence.

    He *is* tall.   She *looks* beautiful.   You *seem* tired.

- **Vowel** The following letters are vowels: *a, e, i, o, u.* *Y* is sometimes considered a vowel (for example, in the word *syllable*).

APPENDIX L

Special Uses of Articles

No Article	Article
Personal names: John Kennedy Michael Jordan	The whole family: the Kennedy the Jordans
Title and name: Queen Elizabeth Pope John Paul	Title without name: the Queen the Pope
Cities, states, countries, continents: Cleveland Ohio Mexico South America	Places that are considered a union: the United States the former Soviet Union the United Kingdom Place names: the _____ of _____ the People's Republic of China the District of Columbia
Mountains: Mount Everest Mount McKinley	Mountain ranges: the Himalayas the Rocky Mountains
Islands: Coney Island Staten Island	Collectives of islands: the Hawaiian Islands the Virgin Islands the Philippines
Lakes: Lake Superior Lake Michigan	Collectives of lakes: the Great Lakes the Finger Lakes

Continued

Beaches:
  Palm Beach
  Pebble Beach
	Rivers, oceans, seas, canals:
	    the Mississippi River
	    the Atlantic Ocean
	    the Dead Sea
	    the Panama Canal

Streets and avenues:
  Madison Avenue
  Wall Street
	Well-known buildings:
	    the Sears Tower
	    the Empire State Building

Parks:
  Central Park
  Hyde Park
	Zoos:
	    the San Diego Zoo
	    the Milwaukee Zoo

Seasons:
  summer   fall
  spring   winter
  Summer is my favorite season.
NOTE: After a preposition, *the* may be used.
  In (the) winter, my car runs badly.
	Deserts:
	    the Mojave Desert
	    the Sahara Desert

Directions:
  north   south
  east   west
	Sections of a piece of land:
	    the Southwest (of the U.S.)
	    the West Side (of New York)

School subjects:
  history
  math
	Unique geographical points:
	    the North Pole
	    the Vatican

Name + *college* or *university*:
  Northwestern University
  Bradford College
	The University (College) of _____ :
	    the University of Michigan
	    the College of DuPage County

Magazines:
  Time
  Sports Illustrated
	Newspapers:
	    the *Tribune*
	    the *Wall Street Journal*

Months and days:
  September
  Monday
	Ships:
	    the *Titanic*
	    the *Queen Elizabeth*

Holidays and dates
(Month + Day):
  Thanksgiving
  Mother's Day
  July 4
	The day of (month):
	    the Fourth of July
	    the fifth of May

Diseases:
  cancer   AIDS
  polio   malaria
	Ailments:
	  a cold a toothache
	  a headache the flu

Games and sports:	Musical instruments, after *play*:
poker	the drums
soccer	the piano
	NOTE: Sometimes *the* is omitted.
	She plays (the) drums.

Languages:	The _____ language:
French	the French language
English	the English language

Last month, year, week, etc. = the one before this one:	The last month, the last year, the last week, etc. = the last in a series:
I forgot to pay my rent last month.	December is the last month of the year.
The teacher gave us a test last week.	Summer vacation begins the last week in May.

In office = in an elected position:	In the office = in a specific room:
The president is in office for four years.	The teacher is in the office.

In back/front:	In the back/the front:
She's in back of the car. 	He's in the back of the bus.

Alphabetical List of Irregular Verb Forms

Base Form	Past Form	Past Participle	Base Form	Past Form	Past Participle
be	was/were	been	bite	bit	bitten
bear	bore	born/borne	bleed	bled	bled
beat	beat	beaten	blow	blew	blown
become	became	become	break	broke	broken
begin	began	begun	breed	bred	bred
bend	bent	bent	bring	brought	brought
bet	bet	bet	broadcast	broadcast	broadcast
bid	bid	bid	build	built	built
bind	bound	bound	burst	burst	burst

Continued

Base Form	Past Form	Past Participle	Base Form	Past Form	Past Participle
buy	bought	bought	hide	hid	hidden
cast	cast	cast	hit	hit	hit
catch	caught	caught	hold	held	held
choose	chose	chosen	hurt	hurt	hurt
cling	clung	clung	keep	kept	kept
come	came	come	know	knew	known
cost	cost	cost	lay	laid	laid
creep	crept	crept	lead	led	led
cut	cut	cut	leave	left	left
deal	dealt	dealt	lend	loaned/lent	loaned/lent
dig	dug	dug	let	let	let
dive	dove/dived	dove/dived	lie	lay	lain
do	did	done	light	lit/lighted	lit/lighted
draw	drew	drawn	lose	lost	lost
drink	drank	drunk	make	made	made
drive	drove	driven	mean	meant	meant
eat	ate	eaten	meet	met	met
fall	fell	fallen	mistake	mistook	mistaken
feed	fed	fed	overcome	overcame	overcome
feel	felt	felt	overdo	overdid	overdone
fight	fought	fought	overtake	overtook	overtaken
find	found	found	overthrow	overthrew	overthrown
fit	fit	fit	pay	paid	paid
flee	fled	fled	plead	pled/pleaded	pled/pleaded
fly	flew	flown	prove	proved	proven/proved
forbid	forbade	forbidden	put	put	put
forget	forgot	forgotten	quit	quit	quit
forgive	forgave	forgiven	read	read	read
freeze	froze	frozen	ride	rode	ridden
get	got	gotten	ring	rang	rung
give	gave	given	rise	rose	risen
go	went	gone	run	ran	run
grind	ground	ground	say	said	said
grow	grew	grown	see	saw	seen
hang	hung	hung[2]	seek	sought	sought
have	had	had	sell	sold	sold
hear	heard	heard	send	sent	sent

[2] *Hanged* is used as the past form to refer to punishment by death. *Hung* is used in other situations: She *hung* the picture on the wall.

Base Form	Past Form	Past Participle	Base Form	Past Form	Past Participle
set	set	set	swing	swung	swung
sew	sewed	sewed/sown	take	took	taken
shake	shook	shaken	teach	taught	taught
shed	shed	shed	tear	tore	torn
shine	shone/shined	shone	tell	told	told
shoot	shot	shot	think	thought	thought
show	showed	shown/showed	throw	threw	thrown
shrink	shrank/shrunk	shrunk/shrunked	understand	understood	understood
shut	shut	shut	uphold	upheld	upheld
sing	sang	sung	upset	upset	upset
sink	sank	sunk	wake	woke	woken
sit	sat	sat	wear	wore	worn
sleep	slept	slept	weave	wove	woven
slide	slid	slid	wed	wedded/wed	wedded/wed
slit	slit	slit	weep	wept	wept
speak	spoke	spoken	win	won	won
speed	sped	sped	wind	wound	wound
spend	spent	spent	withhold	withheld	withheld
spin	spun	spun	withdraw	withdrew	withdrawn
spit	spit	spit	withstand	withstood	withstood
split	split	split	wring	wrung	wrung
spread	spread	spread	write	wrote	written
spring	sprang	sprung			
stand	stood	stood			
steal	stole	stolen			
stick	stuck	stuck			
sting	stung	stung			
stink	stank	stunk			
strike	struck	struck/stricken			
strive	strove	striven			
swear	swore	sworn			
sweep	swept	swept			
swell	swelled	swelled/swollen			
swim	swam	swum			

Note:

The past and past participle of some verbs can end in -ed or -t.

burn	burned or burnt
dream	dreamed or dreamt
kneel	kneeled or knelt
learn	learned or learnt
leap	leaped or leapt
spill	spilled or spilt
spoil	spoiled or spoilt

The United States of America: Major Cities

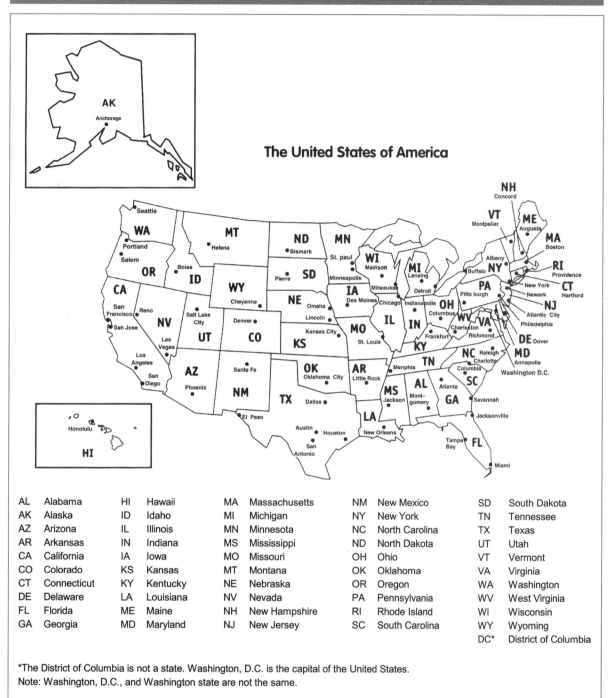

The United States of America

AL	Alabama	HI	Hawaii	MA	Massachusetts	NM	New Mexico	SD	South Dakota
AK	Alaska	ID	Idaho	MI	Michigan	NY	New York	TN	Tennessee
AZ	Arizona	IL	Illinois	MN	Minnesota	NC	North Carolina	TX	Texas
AR	Arkansas	IN	Indiana	MS	Mississippi	ND	North Dakota	UT	Utah
CA	California	IA	Iowa	MO	Missouri	OH	Ohio	VT	Vermont
CO	Colorado	KS	Kansas	MT	Montana	OK	Oklahoma	VA	Virginia
CT	Connecticut	KY	Kentucky	NE	Nebraska	OR	Oregon	WA	Washington
DE	Delaware	LA	Louisiana	NV	Nevada	PA	Pennsylvania	WV	West Virginia
FL	Florida	ME	Maine	NH	New Hampshire	RI	Rhode Island	WI	Wisconsin
GA	Georgia	MD	Maryland	NJ	New Jersey	SC	South Carolina	WY	Wyoming
								DC*	District of Columbia

*The District of Columbia is not a state. Washington, D.C. is the capital of the United States.
Note: Washington, D.C., and Washington state are not the same.

North America

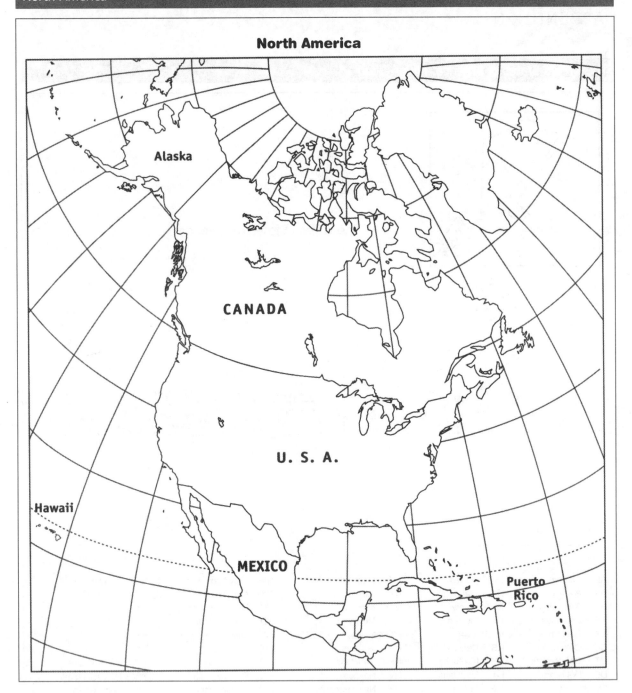

Index

Photo Credits

2, left LWA-Dan Tardif/CORIBS, *right* Bill Truslow/Image Bank/Getty Images; *4,* Mary Kate Denny/Stone/Getty Images; *6,* Ann Marie Weber/Taxi/Getty Images; *14,* Stephen McBrady/PhotoEdit; *18,* Syracuse/Newspaper/David Lassman/The Image Works; *24,* Pam Gardner/Frank Lane Picture Agency/CORBIS; *25,* Michael Simpson/Taxi/Getty Images; *26,* Tim MacPherson/Stone/Getty Images; *43,* Tom and Dee Ann McCarthy/CORBIS; *44,* Ariel Skelley/CORBIS; *45,* Marty Heitner/The Image Works; *49,* Robin Sachs/PhotoEdit; *57,* Bonnie Kaufmann/CORBIS; *64,* Sandra Elbaum; *65,* Ariel Skelley/CORBIS; *68,* David Houser/CORBIS; *70,* Bonnie Kaufmann/CORBIS; *87, top,* Flip Schulke/CORBIS, *bottom,* William Philpott/Reuters/CORBIS; *88,* Bettmann/CORBIS; *89, left,* Duomo/CORBIS, *right,* Frank Trapper/CORBIS; *93,* Thomas Gilbert/AP/Wide World Photos; *97,* Bettmann/CORBIS; *100,* CORBIS; *115,* Rick Gomez/CORBIS; *126,* Steven Rothfeld/Stone/Getty Images; *127,* Brand X Pictures/Getty Images; *129,* Chuck Savage/CORBIS; *134,* Susan van Etten/PhotoEdit; *138,* Tony Freeman/PhotoEdit; *149,* Bob Daemmrich/The Image Works; *150,* Kelley Mooney Photography/CORBIS; *157,* Bruce Ayres/Stone/Getty Images; *166,* Ron Sachs/CORBIS; *168,* Tom Bean/CORBIS; *170,* Marty Heitner/The Image Works; *173,* Sandra Elbaum; *184,* Peter Turnley/CORBIS; *187, left,* Eric K.K. Yu/CORBIS, *right,* Royalty Free/CORBIS; *188,* Robert E. Daemmrich/Stone/Getty Images; *189,* ANA/The Image Works; *194,* Michael Newman/PhotoEdit; *195,* Dana White/PhotoEdit; *211,* CORBIS; *212,* Gail Mooney/CORBIS; *213,* Sandra Elbaum; *218,* Hulton Archive/Getty Images; *227,* Nancy Kaszerman/Zuma/CORBIS; *237,* Michael Newman/PhotoEdit; *239,* Shelley Gazin/The Image Works; *246,* David Young Wolff/PhotoEdit; *249,* Dwayne Newton/PhotoEdit; *252,* Bonnie Kamin/PhotoEdit; *257,* James Marshall/The Image Works; *262,* Willie Hill, Jr./The Image Works; *277,* Dex Images/CORBIS; *278,* Kim Kulshi/CORBIS; *289,* Aleata Evans; *292,* Images.com/CORBIS; *298,* Richard Lord/The Image Works; *315,* Ralf-Finn Hestoft/CORBIS; *316,* Royalty Free/CORBIS; *353,* Randy M. Ury/CORBIS; *354,* Tony Freeman/PhotoEdit; *361,* Aleata Evans; *364,* David Young Wolff/PhotoEdit; *375,* Duomo/CORBIS; *375,* Mike Segar/Reuters/CORBIS; *376,* Reuters/CORBIS; *377,* Reuters/CORBIS; *381,* Reuters/CORBIS; *382,* Reuters/CORBIS; *385,* Bob Daemmrich/The Image Works; *390,* Didrik Johnck/CORBIS; *397, left,* Albert Gea/Reuters/CORBIS, *right,* Robert Galbraith/Reuters/CORBIS; *413,* Jose Luis Pelaez, Inc./CORBIS; *414,* Billy E. Barnes/PhotoEdit; *422,* Michael Newman/PhotoEdit; *423,* Billy Aron/PhotoEdit; *437,* Susan van Etten/PhotoEdit; *438,* David Young Wolff/PhotoEdit; *442,* Mary Kate Denny/PhotoEdit; *444,* Jeff Greenberg/The Image Works; *451,* Michael Newman/PhotoEdit; *451,* Joseph Sohm/Visions of America/CORBIS; *456,* Dana White/PhotoEdit; *459,* David Young Wolff/PhotoEdit; *461,* Michele Birdwell/PhotoEdit.

《英语语境语法》（第四版）系列丛书

尊敬的老师：

　　您好！

　　为了方便您更好地使用本套教材，获得最佳教学效果，我们特向使用该套丛书作为教材的教师赠送 CD-ROM 测试题库和教学录像。如有需要，请完整填写"教师联系表"，免费向出版社索取。

北京大学出版社

✂ -

教师联系表

姓名：	性别：		职务：	职称：
E-mail：	联系电话：		邮政编码：	
供职学校：		所在院系：		
学校地址：				
教学科目与年级：		班级人数：		
通信地址：				

　　填写完毕后，请将此表邮寄或 EMAIL 给我们，我们将为您免费寄送 CD-ROM 测试题库及教学录像，谢谢！

北京市海淀区成府路 205 号
北京大学出版社外语编辑部负责人
邮政编码：100871
电子邮箱：zbing@pup.pku.edu.cn

邮购部电话：010-62534449
市场营销部电话：010-62750672
外语编辑部电话：010-62765014

西方语言学原版影印系列丛书

北京大学出版社

邮 购 部 电话：010-62534449　联系人:孙万娟

市场营销部电话：010-62750672

外语编辑部电话：010-62767315　62767347